D1000881

Modern structural analysis

Modelling process and guidance

Iain A. MacLeod

DISCARD

Thomas Telford

Published by Thomas Telford Publishing, Thomas Telford Ltd, 1 Heron Quay, London E14 4JD.
www.thomastelford.com

Distributors for Thomas Telford books are
USA: ASCE Press, 1801 Alexander Bell Drive, Reston, VA 20191-4400, USA
Japan: Maruzen Co. Ltd, Book Department, 3–10 Nihonbashi 2-chome, Chuo-ku, Tokyo 103
Australia: DA Books and Journals, 648 Whitehorse Road, Mitcham 3132, Victoria

First published 2005

A catalogue record for this book is available from the British Library

ISBN: 0 7277 3279 X

© Thomas Telford Limited 2005

All rights, including translation, reserved. Except as permitted by the Copyright, Designs and Patents
Act 1988, no part of this publication may be reproduced, stored in a retrieval system or transmitted in
any form or by any means, electronic, mechanical, photocopying or otherwise, without the prior written
permission of the Publishing Director, Thomas Telford Publishing, Thomas Telford Ltd, 1 Heron
Quay, London E14 4JD.

This book is published on the understanding that the author is solely responsible for the statements
made and opinions expressed in it and that its publication does not necessarily imply that such
statements and/or opinions are or reflect the views or opinions of the publishers. While every effort
has been made to ensure that the statements made and the opinions expressed in this publication provide
a safe and accurate guide, no liability or responsibility can be accepted in this respect by the author or
publishers.

Typeset by Academic + Technical, Bristol
Printed and bound in Great Britain by MPG Books, Bodmin

TA
645
.M18
2005

Acknowledgements

This book follows on from *Analytical modelling of structural systems* published in 1990. I was involved in a working group of the Institution of Structural Engineers which resulted in the 2002 publication of the booklet *The use of computers for engineering calculations.* A number of ideas about modelling process which I have used in the book arose from the work of the group and I acknowledge with thanks the contributions of Andrew Bond, Peter Gardner, Peter Harris, Bill Harvey, Nigel Knowles and Brain Neale to these ideas.

I am specially grateful to Sam Thorburn, Yaqub Rafiq and Steven McKerlie who read a draft of the book and provided me with many useful suggestions.

I record my thanks to the following people for advice and information on the production of this book: Kamal Badrah, Callum Bennett, Prabakhara Bhatt, Roy Cairns, Andrew Clark, Graeme Harley, Paul Lyons, John Morrison, Matthew Petticrew, Ian Salisbury, David Scott, Richard Wood, Howard Wright and Karoly Zalka.

Finally my thanks to Barbara, Mairi, Alastair and Iseabail for their love and support.

Foreword

This interesting book promotes a new way of looking at structural analysis. It suggests that the ability to work with the model (as distinct from the solution process) is a primary issue which should be formally addressed in practice and in education. The content is focused on modelling issues and I know of no other text which does this so comprehensively.

The early chapters contain much advice necessary to help the reader establish how to formulate a numerical model that might be capable of simulating the performance of the actual structural system under investigation. The later chapters include a good outline of the issues involved in modelling of structures using finite elements. The two case studies given at the end of the book are a good device to put the excellent advice given in the earlier sections into some perspective for the reader.

I found it most useful to have in the same book a reminder of the theoretical basis of the full range of finite element types and a sound method as to how to employ analysis as a reflective tool towards a better understanding of structural behaviour. The rigorous treatment for the process of validation of a model is most enlightening as is that outlined for verification of the results. After all, the iterative process of model validation and output verification are the main activities for gaining a true understanding of structural behaviour.

My own experience working with Buro Happold tells me that robust structural design requires the willingness to develop an understanding of structural behaviour with a questioning mind. In most consulting offices, current practice is to undertake this using finite element models of increasing complexity as understanding of the problem at hand grows. Iain MacLeod describes clearly how to build up this understanding using sensitivity analysis and simplified loadings to test validity against expectations from parallel calculation and modelling experiences. It is argued that risk will be reduced in practice if there is a rigorous analytical process that reflects the realities of current engineering practice in most offices.

Most structures are of a reasonably conventional type and use well tried framing systems. Substantial experience already exists on their likely performance so hand calculations based on structural theory can be done to initiate formulation of the model or to act as a check on the results. However, even advanced classical methods struggle to model the sophistication of load paths in redundant or non-linear structures where individual stiffness, material response and definition of restraint determines structural performance. In this case, I have found that comparison of the output of simplified analytical results with physical models very useful as an addition to classical calculation – as advocated in the second chapter.

The book is thus both a useful reference for the practitioner and a comprehensive learning guide for the student. It builds on the publication by the Institution of Structural Engineers *Guidelines for the Use of Computers for Engineering Calculation* published in 2002. Its carefully constructed content successfully redresses the imbalance in risk between the finite element process based around generally determinate calculation output that has itself been derived from a possibly non-determinate understanding of the actual modelling process. In the Introduction, the author suggests that all structural engineers and all civil engineers who use structural analysis will find the contents of the book to be useful. I think that he is right.

Michael Dickson FIStructE
Director, Design and Technology Board, Buro Happold
President, Institution of Structural Engineers 2005–06

Contents

1 Introduction

1.1 Scope and definitions

In this context *structural analysis* is the use of mathematical models of structural engineering systems, such as building frames and bridges, to predict stresses, deformations, etc., which result from loading.

Structural analysis modelling is the process of creating and using structural analysis models.

This book:

- discusses basic principles for structural analysis modelling
- promotes the use of a process which helps to minimise the inherent risks
- discusses detailed issues for basic and intermediate level analysis modelling – with emphasis on the modelling of skeletal frames.

It is expected that both practitioners and students in structural engineering and in civil engineering will find the contents to be useful.

1.2 Why 'modern' structural analysis?

Two basic processes in structural analysis are:

- a model development process – essentially a synthesis process
- a solution process – the analysis component.

Traditionally, the potential scope of a structural analysis model was so limited that the model development process was a minor issue; the major issue was how to achieve a solution. Now, the potential wide scope of models, and the fact that the solution process is simple – the computer does the work – makes the model development process the major issue.

The result of the solution process is determinate, i.e. there is a unique solution for a given model. In such a situation the difference between the results and the model is *error*. The result of the model development process is, however, non-determinate (there may be several valid models) and the difference between the system being modelled and the model is characterised by *uncertainty*.

Therefore, the emphasis for the structural engineer has changed radically from analysis to model synthesis, from contexts where the outcomes are unique to contexts where uncertainty plays a dominant role. This is the reality, but our perception of the situation has been slow to change. The paradigm shift has not been identified in education and hence it is not well understood in practice.

This textbook addresses this modern reality of the subject.

1.3   Issues for practice

Computer solutions in structural analysis are much more efficient than hand solutions, but it is not uncommon for senior engineers to require that new graduates start by doing hand solutions so as to develop understanding of behaviour. The principle that doing hand calculations is essential to proper understanding is deeply rooted in the conventional wisdom of structural engineering. However, the degree of understanding gained from *doing* hand calculations is normally overestimated. Reflective consideration of results is a more fruitful source of understanding than arithmetical processing.

Use of the modelling process described in Chapter 3 and the methods for understanding behaviour described in Section 2.4 stimulate such reflection and understanding. They have much greater potential for developing understanding of behaviour than do hand solutions.

Risk is defined as the combination of the likelihood and the consequences of an event which can cause harm. The likelihood of occurrence of a disaster due to structural analysis modelling is low but the potential consequences from such an event are very severe. The highest recorded number of lives lost in a structural failure in the UK (75) resulted from the collapse of the Tay Bridge in 1887. This pre-computer disaster was mainly due to inadequate modelling (Section 3.8.1). The collapse of the roof of the Hartford Center, Connecticut, in 1979 (Section 3.8.2) was due to faulty modelling using computer solutions. It occurred when the building was empty, but had it happened when the building was in use the loss of life would have been great. The risk of a disaster causing serious harm due to inadequate modelling cannot be eliminated; it can only be minimised. But to be minimised, the modelling process discussed in Chapter 3 needs to be formally adopted.

1.4   Issues for education

Table 1.1 assesses the relative importance of fundamental abilities for structural analysis needed in practice as compared with that given in education.

In education, the importance of basic mechanics is accepted but the concept of modelling process is not part of the conventional wisdom. Although teachers might claim that the learning programmes that they use do help to promote

Table 1.1 Abilities of structural analysis

Ability	Relative importance	
	needed in practice	given in education
Understand basic principles of structural mechanics, including equilibrium, compatibility and force-deformation relationships	High	High
Use the modelling process	High	Neglected
Develop understanding of the behaviour of the system being modelled	High	Low
Understand the (computer) solution process	Low	High

understanding of structural behaviour, it still tends to be assumed (erroneously as explained in Section 1.3) that this is mainly achieved by doing hand calculations. Important strategies for developing such understanding are ignored. A significant amount of learning time is normally devoted to understanding the computer solution process, giving only secondary advantage. It is therefore evident that education for structural analysis is significantly out of balance with the needs of practice.

To suggest that understanding the solution process is of secondary importance is often viewed as heretical. But whereas one cannot use modelling software effectively without competence in mechanics, modelling process and understanding of behaviour (first three abilities listed in Table 1.1), one could get by with a very limited amount of knowledge about the solution process. It is not suggested that the solution process should be neglected in education; rather that it need not be treated as being of primary importance.

Because emphasis is placed on the solution process in the teaching of structural analysis, learners tend to develop the attitude that the process is determinate and with unique answers. But, as pointed out in Section 1.2, the model development process is non-determinate and hence the overall process is non-determinate.

Implicit in the discussion on modelling process in this book is the need for education to take account of the paradigm shift outlined in Section 1.2 by:

- addressing the model development process, with emphasis on uncertainty
- developing a reflective approach to modelling, where consideration of validation of the model, verification of results, sensitivity analysis and a purposeful approach to developing understanding of behaviour are main activities
- providing a good grounding in basic mechanics, but treating computer solution methods as being of secondary importance.

1.4.1 The wider context

If a paradigm shift is needed in education for structural analysis then a similar shift is needed for the teaching of modelling in other domains and disciplines – hydraulic modelling, geotechnical modelling, etc. Engineers were concerned that the introduction of computers would result in dumbing down of engineering. 'Monkeys' could be trained to hit the right keys. While there is a significant danger that software can be used in the absence of competence, the reality of computer use for modelling is that it has made the work more intellectually challenging. To operate successfully in environments of significant uncertainty requires intellectual power of the highest order. This is the realm of modern structural analysis – the realm of the professional engineer.

1.5 Finite elements

The main type of mathematical model used in practice for structural analysis, and discussed in this book, is an *element* model: where the structure is divided into elements.

Prior to the advent of electronic computers in the 1950s structural engineers were only able to obtain solutions for simple frame element models, but a wide range of *finite element* types is now available (see Chapter 4).

1.6   Accuracy of the information provided in the text

While the text has been prepared carefully to avoid errors there can be no guarantee that there are none. Also, because the main issue in establishing a model is uncertainty rather than error, the text avoids being prescriptive about modelling issues.

An important concept in this book is *validation information*, which is information about assumptions and their validity. There is no claim of generality for the validation information given in the text. The lists of assumptions may not be complete, and the discussion of validity may not address all the relevant issues. They are there for guidance.

1.7   Website

Supplementary information on modelling, including examples that relate directly to items in this book, can be found at: www.imacleod.com/msa.

2 Basic principles

This chapter sets out a number of general principles that underpin structural analysis modelling work.

2.1 Managing the analysis process

2.1.1 Quality management system

If you are doing structural analysis you need to have a system for getting it right – a quality management system. Information for developing such a system can be gained from Chapters 2 and 3 of this book and from IStructE (2002) and NAFEMS (1995, 1999).

2.1.2 Use the modelling process

The central feature of the quality management system is the definition of the process to be used. The process activities discussed in Chapter 3 should form the basis of the modelling process.

2.1.3 Competence

A main feature of a quality plan is the need to ensure that those who are involved in the process have the necessary ability and training to do the work.

2.2 Modelling principles

2.2.1 Use the simplest practical model

It is good practice to keep the model at as simple a level as is practical.

An example of an over-complex model is that used for the design of the Sleipner oil platform (Section 3.8.3). The use of shell elements to model the cell walls would probably have been better than using a single thickness of three-dimensional (3D) brick elements.

Refinement of a model may not deliver improvements if fundamental issues are not addressed. For example:

- in Section 9.5 it is noted that post-yield behaviour is fundamental for seismic design of conventional buildings and refinement of elastic models may not be advantageous
- in Section 8.3.1 it is noted that for analysis for differential settlement of a building the variation of soil stiffness over the area of the building, which is seldom addressed in modelling, may be the dominant issue.

This may be stated in the form of a *model refinement principle*:

> If you are using a model in which a feature that dominates the behaviour is neglected, there may be no value in making refinements to the modelling of the features of the behaviour that are included in the model.

2.2.2 Estimate results before you analyse

Some designers advise that if you do not have a good estimate of the likely values of the results before you do an analysis then you are not competent for the work. This is an extreme view since with unusual structures a main objective of the analysis is to develop the ability to make such predictions. However, it is always a good idea to estimate results at the outset and to check them later with those from the analysis model. If they correspond, then possibilities are that:

- you do have a good understanding of the system
- you made a lucky guess
- both your estimate and the analysis results are wrong (false correlation – see Section 3.6.3).

If they do not correspond and you can identify the reason, then you are improving your understanding of behaviour.

2.2.3 Increment the complexity

If you decide to move into an area of analysis that is unfamiliar, build up experience by starting with simple (smaller) elastic models and loadcases for which solutions are known (if practical). If using non-linear analysis, start with an elastic model, then move into separate non-linear material and non-linear geometry models and then combine them. At each stage review the results to assess whether or not they are acceptable.

2.2.4 When you get results, assume that they may be errors

Treat results in the opposite way to an accused person in law. Analysis results should be treated as being guilty (of having errors) until they have passed rigorous scrutiny. It is not uncommon to find errors after all checks appear to have been satisfactorily completed.

2.2.5 Troubleshooting for errors

It is common when running analysis software either to get no results or to get results that do not make sense. Even if they make sense this does not mean that they are correct. It is said that steps in structural analysis modelling are to:

- obtain results
- obtain realistic results
- obtain correct results.

Fundamental sources of error are:

- *Data errors* – this the most likely source of errors.

- *Software errors* – these are uncommon when using a well-established system but cannot be ruled out (see Section 3.5.2).
- *Hardware errors* – these can happen but are a very unlikely last resort possibility. Hardware errors normally result in a 'crash', but the author has been in the situation where there was an error in the double-precision arithmetic which only affected the less-significant ends of the arithmetic operations.

When an error is either obvious or suspected, the following actions can be taken.

- Check the data – obvious first action. Two common sources of error here are incorrect data values and misinterpretation of the data format.
- Check the output for solution error warnings.
- Use a checking model (Section 3.6.3).
- Make changes to the model which will help to draw out the problem. A good strategy is to review the results with the simplest practical loadcase – for example, one which corresponds to a realistic checking model.
- Set up a simpler version of the model and compare the results.
- For important projects, re-establish the model using different software.

2.2.6 Relationship between the analysis model and the design code of practice

The validity of an analysis model needs to be considered in relation to the code of practice to be used for sizing the members. Structural codes of practice are based on standard practice for specific structural types. If the structure is within the scope of the design code of practice, then validation of the analysis model conventionally used in relation to that code can be brief. For example, if you are designing a steel portal frame structure then the approach to analysis recommended in the code may be readily accepted. However, if the type of structure to be designed is to some degree outwith the scope of codes of practice then much more careful validation may be needed – see Section 2.2.7.

The follow-through principle

The basis of this principle is that when assumptions are made for the analysis model they should be incorporated into the member sizing processes. For example, for a triangulated frame, if bending of the members is taken into account in the analysis model then it also has to be taken into account in the member sizing process (Fig. 2.1). All the internal forces from an analysis model should be considered in the member sizing process. If some are ignored then the internal forces used for design will not be in equilibrium with the applied load and the

Figure 2.1 Truss model using beam elements with moment connections – bending must be considered in the member sizing process.

fundamental condition for the lower bound theorem (Section 2.3.3) to be valid will have been violated.

2.2.7   Case study – the Ronan Point collapse

The Ronan Point building was a 22-storey residential building in Canning Town, London, UK. The vertical loadbearing structure consisted of storey-height precast concrete panels (no columns); the floors were room-sized precast panels (no beams). It was of typical 'large panel' construction, which was widely used in the UK and in Europe in the 1960s as a strategy for providing low-cost housing (although it tended not to deliver low cost). It was a reinforced-concrete building and UK structural designers used the then current *CP114 Code of Practice for the Structural Design of Reinforced Concrete Buildings* for such buildings. The title of the code seemed to fit the situation, but it had been devised mainly for the design of cast-in-situ beam-and-column frames. It was not meant to be used for large-panel construction and therefore did not address the crucial issue of connections between large panels.

Early on the morning of 16 May 1968 (Griffiths *et al.* 1968) a woman in an eighteenth-floor apartment of the building struck a match to light a cooker. This initiated a major gas explosion (the presence of gas in the room was due to a faulty pipe fitting), causing the external walls of her flat to blow out and the then unsupported structure above the explosion to collapse. The debris loading then took away the panels in the corner of the building below the explosion. Because of the early hour only four people were killed.

This was one of the most important structural incidents of the twentieth century. The designers were eventually held liable.

The collapse was not due to major errors in a structural analysis, such as in the Sleipner platform collapse (Section 3.8.3). But if structural analysis is thought of as the process of predicting the behaviour of structures, this was a modelling failure. The designers did not imagine a 'house of cards' analogy for the system. The fundamental lesson for structural analysis is that if you justify the analysis model on the basis of a code of practice, then you must be sure that the system that you are designing is within the scope of the code.

2.3   Principles in the use of structural mechanics

2.3.1   Local and resultant stresses – the St Venant principle

This is an important principle in relation to the level of detail that should be used in an analysis. The St Venant principle (not to be confused with St Venant torsion – Section 5.4.2) can be stated as follows:

> Forces applied at one part of an elastic structure will induce stresses which, except in a region close to that part, will depend almost entirely on their resultant action and very little on their distribution.

Local stresses are caused by loading applied in the vicinity of the stresses. *Resultant stresses* are caused by resultant force actions, for example in a beam element the bending moment, the axial force and the torque are resultant actions.

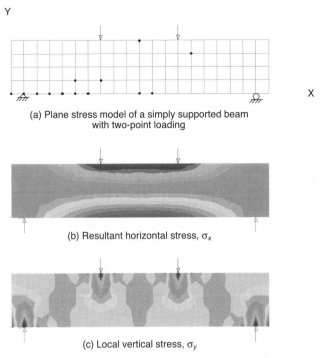

Figure 2.2 Local and resultant stresses.

The St Venant principle has the following consequences.

- The effect of local stresses tends to become negligible at distances twice the length over which the stress concentration acts.
- When calculating resultant stresses, the force actions can be replaced by static equivalents. This is not so for local stresses.

Figure 2.2(a) shows a simply supported beam modelled as plane stress elements with applied two-point loading. Figure 2.2(b) shows the horizontal direct stress, σ_x. These are resultant stresses due to the bending moment, but there are some local effects in the vicinity of the point loads. Figure 2.2(c) is a plot of σ_y. These are entirely local stresses which would be not be taken into account if the system was modelled using beam elements. (See also the local stresses for the axially loaded plate in Fig. 5.9.)

It is important to distinguish between local and resultant stresses in modelling. A good approach is to have an overall model for the resultant stresses and to treat local stresses using models which incorporate only local features.

2.3.2 Principle of superposition
Basis of the principle
In a linear elastic analysis, the order of applying loads does not affect the final results. Separate loadcases can be added to give the final result. For example,

two loads W_1 and W_2 are to be applied to an linear elastic structure. If at any point i

- the deformations and internal actions due to W_1 are Δ_1 and p_1, and
- the deformations and internal actions due to W_2 are Δ_2 and p_2

then the deformations and internal actions due to the combined effect of W_1 and W_2 will be

$$\Delta_{\text{combined}} = \Delta_1 + \Delta_2 \qquad\qquad (2.1)$$

$$p_{\text{combined}} = p_1 + p_2 \qquad\qquad (2.2)$$

Validation information
The principle of superposition only holds for linear elastic structures. With non-linear material properties (Section 7.3) or non-linear geometry (Chapter 10) the analysis process must take account of the realistic order of the loading, i.e. loads must be applied in the order in which they would occur in the real situation.

2.3.3 Lower bound theorem in plasticity
The lower bound theorem can be stated as follows:

> If a set of internal forces is identified which is in equilibrium with the applied load on a structure and the yield criterion is not anywhere exceeded then the corresponding applied load is less than or equal to the collapse load. (Moy 1996)

It is also a requirement that the ductility (Section 7.3.1) is everywhere adequate to sustain the yield conditions. For example, suppose that a frame which can form plastic hinges is to be designed for an ultimate load of W_u. A set of bending moments for the frame is devised which are in equilibrium with W_u. Each member of the frame is sized such that it will form a plastic hinge at the specified bending moment. When the real load is applied, hinges will form in some order but, *as long as they continue to provide their yield moment of resistance,* when the last plastic hinge occurs to form a mechanism, the applied load – W_c – will be greater than (or equal to) W_u, i.e. the design load will be less than the real collapse load and so the design will be safe.

A main criterion for validity of the theorem is the clause in italics in the previous paragraph. Once a section has reached its yield moment it will provide that moment only up to its ductility capacity. Ductility is never unlimited; if part of the structure goes beyond its ductility capacity before the collapse load is reached the lower bound theorem will not be valid.

The lower bound theorem is used to justify the use of elastic analysis for design for ultimate load. If the factored loads from an elastic analysis are used to size all the members of a frame, then all sections will reach their yield load at the ultimate load and the design ultimate load will be the collapse load. This is more practical for concrete structures than for steel structures, but even then it is likely that the detailing will be standardised such that the section strengths are in many situations greater than that required by the elastic analysis. This makes the design more conservative.

The lower bound theorem is also used to justify the neglect of some of the internal actions in the member sizing process. For example, in a triangulated frame, if the analysis model does not include bending (i.e. only bar elements are used – Section 5.5.1), the axial forces will provide a set of internal actions in equilibrium with the applied load which can be used to size the members. The yield criterion (but not necessarily the ductility requirement) will be satisfied. It seems possible, however, that neglecting structural actions in this way could, in certain circumstances, prevent the assumed ultimate strength from being realised.

2.4   Understanding structural behaviour

2.4.1   General

An important purpose of structural analysis is to promote understanding of the behaviour of the system. In this section, strategies for the development of such understanding are discussed.

The development of understanding of behaviour is an essential activity in structural design, especially when anything non-standard is being developed. Contrary to some views, the use of computers has high potential to support improvement in understanding. But understanding comes from within. Structural engineers need to drive themselves towards it. They need to be students of behaviour (in the sense of 'student' as one who studies). They need to use every opportunity and use every sense to get 'inside' the systems for which they are responsible.

2.4.2   Model validation

In order to validate a model (Section 3.4) it is necessary to understand the assumptions made in creating it and to relate these to the behaviour of the system being modelled. Developing ability in validation improves understanding of behaviour and vice versa.

2.4.3   Results verification and checking models

Reflective assessment of results helps to verify the model (i.e. to ensure that it has been properly implemented) and to improve understanding of behaviour.

A main activity in results verification is the use of checking models (Section 3.6.3) which can improve understanding in the following ways.

- Correlation between the main model and the checking model strengthens the levels of understanding.
- Explaining the reasons for a faulty checking model creates understanding.

2.4.4   Sensitivity analysis

Features of sensitivity analysis

Sensitivity analysis is where the effects of different values for features or parameters are investigated. This can prove very useful in promoting better understanding of the behaviour of systems being modelled. For example, if the results are not sensitive to the value of a particular parameter this can help to validate the model and provide information which can help in the design. If the results are sensitive to a parameter then special validation action may be needed.

The need for sensitivity analysis is a function of the degree of uncertainty about the behaviour. When working on an unusual system, sensitivity analysis may be essential. The following issues are relevant to sensitivity analysis.

- A useful strategy is to work from a *reference model*, changing one variable at a time and reverting to the reference model after each change. As one gains understanding it may be better to change the reference model, but if the changes are compounded it becomes difficult to make sensible comparisons. (See the example of a sensitivity analysis in the modelling review in Table 12.6.)
- Make comparisons with *indicative parameters*, i.e. parameters which tend to exemplify the behaviour. Typical indicative parameters are:
  o maximum deflection in the direction of the main loading
  o the deflection in the line of a single point load used as a checking loadcase
  o the maximum bending moment, shear force, axial force or torque in the system
  o the value of the lowest natural frequency.
- In the report for the sensitivity analysis, show the corresponding results as close to each other as practicable – on the same table or on the same graph.
- Make the results non-dimensional. Quote percentage changes or ratios of the values to those from the reference model (see, for example, Section 12.2.6). Make the independent variables non-dimensional if practical.

Case study – sensitivity analysis of shear walls with openings
Figure 2.3 is an example of my early experience of learning from a sensitivity analysis (MacLeod 1967). I first took a wall with openings at each storey level (Type A) and then a wall with openings at alternate storey levels (Type B). Then for each type I determined two subtypes: with centred openings (Types A1 and B1), and with off-centre openings (Types A2 and B2).

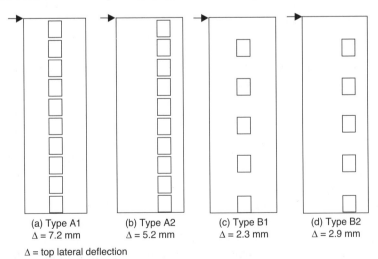

(a) Type A1   (b) Type A2   (c) Type B1   (d) Type B2
Δ = 7.2 mm   Δ = 5.2 mm   Δ = 2.3 mm   Δ = 2.9 mm

Δ = top lateral deflection

Figure 2.3 Shear walls with openings.

For the Type A situation the stiffness increased when the openings were moved off-centre, but the opposite was the case for Type B. I searched for errors but found none. Eventually I realised that for the Type A walls the sum of the I values, and hence the stiffnesses of the wall sections (the parts of the wall on either side of the openings), was greater for the off-centre openings situation than for the central openings. Because of the weak interaction between the walls via the connecting beams, this effect dominated the behaviour and caused the deflection to decrease (stiffness to increase when the openings move off-centre). For the Type B walls, which behave more like a single wall with some cut-outs (rather than two separate wall sections), the off-centre openings are taking material from a more highly stressed part of the wall and hence the deflection increases.

The results were correct, but my interpretation of behaviour was wrong. The sensitivity analysis led to deeper understanding of how walls with openings behave.

2.4.5   Solution comparisons

Making comparisons between results for different solution models can provide important information for understanding and for validation.

When comparing solutions by different methods it is important to work with a *reference solution*. The following types of reference solution may be used.

- *Exact solution* – in principle, if you define a domain with a governing differential equation, boundary conditions and loading there will be an 'exact' solution, say for the deflection in the line of the load, which will be independent of the means of obtaining the solution. In some situations the 'exact' solution may be obtained by summation of a series (e.g. a Fourier series) and can therefore be defined to a desired degree of accuracy (by including as many terms in the series as necessary). Thus we use the term *exact* to mean 'defined to a known high level of accuracy'. Such solutions tend only to be available for situations with limited complexity and applicability.
- *Benchmark solution* – that which has the best known accuracy. It could be an exact solution or could be a result from a fine mesh of finite elements (different from an exact solution because the accuracy may not be known, other than that it is high) – see Section 4.3.4.
- *Arbitrary reference solution* – one against which results are compared; not necessarily the 'exact' solution as in Section 4.4.4.

When comparing results it is important not to jump quickly to conclusions. Validation information about the methods needs to be considered. For example, Fig. 2.4 shows a set of curves for three different models, A, B and D, compared with the results of physical testing, line C. The natural reaction would be to say that line B is the best solution. However, model B may just happen to give a good correlation in this particular situation because of compensating assumptions and may not be better than A or D in all cases.

An example of compensating assumptions for the stiffness of a frame would be if the supports are assumed to be pinned (less stiff than the real situation) and the beam-to-column connections are assumed to be fully fixed (more stiff than the real situation).

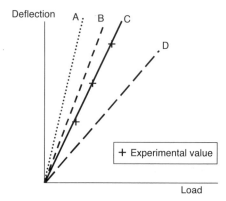

Figure 2.4 Solution comparisons.

2.4.6 Convergence analysis
This is used for assessing meshing strategies for finite element models – see, for example, Section 4.3.2. The use of benchmark solutions, as described in Section 2.4.5, is also important for convergence studies.

2.4.7 Identify patterns
Identifying patterns is a main issue in understanding behaviour. It is worthwhile to collect patterns of behaviour in relation to internal force actions, stresses, deformations, etc. See, for example, Figs 5.2, 5.5 and 5.8.

2.4.8 Mathematics
A mathematical relationship, for example a constitutive relationship (the relationship between stress and strain), represents a model of behaviour. Therefore, the mathematics can be a window to help to view the real behaviour. The mathematics may seem complex when first used, but familiarity develops and meaning emerges.

2.4.9 Physical modelling and testing
The observance of physical behaviour is one of the most important strategies for the development of understanding. Just playing around with simple physical models can be most worthwhile. Opportunities for observing the measured effect of loading on structural systems are not common but should be sought. Comparisons between analysis and test results are, of course, fundamentally important for innovative situations. The commissioning of tests in innovative situations should be given serious consideration.

3   The modelling process

3.1   Overview of the modelling process

3.1.1   General

The process discussed here is basically that advocated in other publications, for example IstructE (2002), MacLeod (1995), NAFEMS (1995, 1999) and ISO 9001 (2000). It tends to be used in a formal way by those who specialise in analysis modelling, and in a tacit way by many practitioners.

In order to reduce risk in analysis modelling a formal modelling process should always be adopted. By *formal* is meant that a written record of the activities of the process should be produced. Reasons for formalising the process include the following:

- it helps to minimise the risks in the use of structural analysis
- it helps to avoid omission of important activities.

Making the process formal provides evidence of the use of good practice should the adequacy of the modelling work be later questioned.

The process described here is for structural analysis contexts but it is directly relevant to any analysis modelling situation (e.g. geotechnical models, hydraulic models, etc.) and can be adapted to other types of model, such as physical models, etc.

3.1.2   Representations of the modelling process

A *determinate* process is one for which there is a unique result. Having decided on a structural analysis model, the solution process provides an unique set of results and hence is determinate. The only part of the modelling process which is determinate is the solution process. A *non-determinate* process does not have a unique solution. All the other activities of the modelling process have non-determinate outcomes and therefore the overall modelling process is non-determinate.

Figure 3.1 and Table 3.1 give different views of the modelling process. Figure 3.1 is a flow diagram of the modelling process: the boxes represent outcomes (no fill for the box) or subprocesses (grey fill for the box). Table 3.1 is another view of the process, one which emphasises the need for acceptance criteria at each stage.

Although these views can be interpreted as implying a linear implementation, the real process is likely to involve much looping back to previous stages – it will not normally be linear. It is not possible to model such non-linearity and therefore Fig. 3.1 and Table 3.1 are not strictly definitions of process but rather are a list (Fig. 3.1) and a matrix (Table 3.1) of activities and outcomes set out in an order in which they normally first occur.

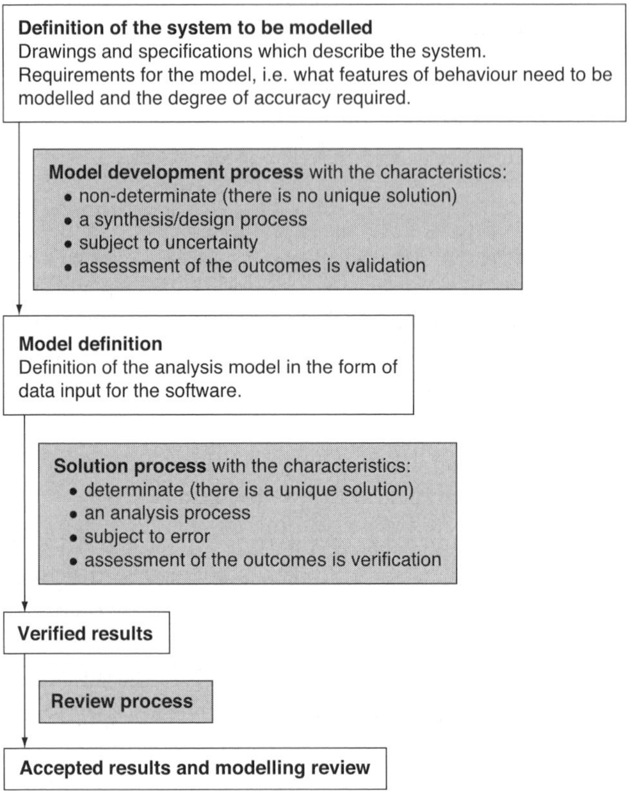

Definition of the system to be modelled
Drawings and specifications which describe the system.
Requirements for the model, i.e. what features of behaviour need to be
modelled and the degree of accuracy required.

Model development process with the characteristics:
- non-determinate (there is no unique solution)
- a synthesis/design process
- subject to uncertainty
- assessment of the outcomes is validation

Model definition
Definition of the analysis model in the form of
data input for the software.

Solution process with the characteristics:
- determinate (there is a unique solution)
- an analysis process
- subject to error
- assessment of the outcomes is verification

Verified results

Review process

Accepted results and modelling review

Figure 3.1 The modelling process.

Table 3.1 Modelling process matrix

	A Model development	B Acceptance criteria	C Model assurance
1 Input	Define the system to be modelled		
2 Analysis model	Define the analysis model	Define acceptance criteria	Validate the analysis model
3 Software	Select suitable software	Define acceptance criteria	Software validation and verification
4 Results	Perform calculations to get results	Define acceptance criteria	Results verification
5 Review		Define overall acceptance criteria	Carry our sensitivity analysis Accept or reject the overall solution Produce modelling review document
6 Output	Define the results to be used for design		

Table 3.2 Modelling activities checklist

1	Define the requirements
2	Validate the model
3	Verify the results
4	Review the outcomes

The process activities set out in Fig. 3.1 and Table 3.1 are normally used by those who do structural analysis. What is often not standard is the treatment of some of the activities in a formal way. In particular, the activities listed in Table 3.2 are often not given enough attention or adequately recorded. Attention to these activities can significantly reduce the risk inherent in structural analysis.

3.1.3 Validation and verification

The following definitions are used in this text (IStructE 2002).

- *Validation* is the consideration of whether or not a process is suited to its purpose. The fundamental question in validation is: is the process capable of satisfying the requirements? – or alternatively: is it the right process?
- *Verification* is the consideration of the question: has the process been implemented correctly? – or alternatively: is the process right?

These definitions are in general agreement with those given in ISO 9001 (2000).

3.1.4 Error and uncertainty

In a modelling process, it is necessary to work with the deviations between the benchmark value of a variable and the value that you have. The *benchmark value* is the desired value of the variable. This leads to the following view of the difference between error and uncertainty.

- *Error* is deviation where the benchmark value is 'exact' – see Section 2.4.5. It is the result of a determinate process. For example, a set of simultaneous equations normally has a potentially exact solution (although real solutions are always approximations). Similarly, the value of π is potentially exact (although there will always be an error in stating it).
- *Uncertainty* is the situation where there is no unique result against which given values can be compared. The outcomes from a non-determinate process are subject to uncertainty, as are the values of material constants. For example, there is no unique value for the value of Young's modulus of concrete (Section 7.2.4); the value depends on how it is measured, and even if the same method is used each time there will be differences in the results for every measurement.

In verification, error tends to be the main consideration, and in validation, uncertainty tends to dominate. Appreciation of the difference between error and uncertainty is important because the tolerance in acceptability is likely to be much greater for uncertainty than for error, as shown in the following examples.

- In defining stiffness for a soil, a deviation (uncertainty) of 10% could be satisfactory.
- In the solution of the system equations in a finite element model, an error check for equilibrium or symmetry should compare up to the last significant figures in the output value. Normal double precision arithmetic for finite element solutions gives 13 significant figures, so the sought accuracy is of the order of 10^{-12} – see example in Section 12.1.6.

3.2   Defining the system to be modelled

The definition of the system to be modelled is sometimes called the *engineering model* (IStructE 2002). Items to be considered include the following:

- *Portrayal of the engineering system to be modelled* – this would be mainly in the form of drawings, sketches and specifications.
- *Requirements of the model* – it is essential to define the outcomes that are required from the modelling activity. Typical objectives of modelling are to predict:
  - stresses or stress resultants
  - failure conditions
  - short-term deformations
  - long-term deformations
  - instability conditions
  - dynamic characteristics.

One of the requirements should be a statement of the desired accuracy of the results. This will depend on the context and, especially, on the degree of risk involved, both with respect to the consequences of failure and to the degree of innovation involved.

3.3   The model development process

3.3.1   Conceptual and computational models

The analysis model is the mathematical representation of the system. It has two components (IStructE 2002).

- The *conceptual model* is defined in terms of material behaviour, loading, boundary conditions, etc. For example, in the analysis of a floor slab the conceptual model could involve linear elastic material behaviour, thin plate bending theory and point supports.
- The *computational model* incorporates the means of achieving a solution. In the case of the floor slab model, the computational model could be based on a specific plate bending finite element mesh (Section 6.3.4) or a grillage model (Section 6.3.6). In some cases the boundary conditions may be part of the computational model; for example, an elastic half-space conceptual model can be reduced to a finite size in the computational model by imposing boundary conditions – see Fig. 8.9. In some situations, for example for elastic frame analysis, computational modelling issues may seldom need to be addressed.

Table 3.3 Modelling issues

Generic issue	Chapter reference
Element type and mesh	4, 5, 6
Material type	7
Supports	8
Connections	5.6
Loading	9
Non-linear geometry	10
Dynamics	11

Defining the conceptual model

A useful strategy at the early part of modelling is to draw up an *issue list*, which is used to help in making decisions about the model. In this context an 'issue' is any feature or factor which may need to be considered in relation to the model. The issue list may be used later in the validation process. Table 3.3 gives a list of generic modelling issues (on which Chapters 4 to 11 of this book are based) which must be taken into account when creating the model.

The conceptual model is normally defined by drawings plus material and geometric data, loading, etc.

Defining the computational model

For element models, the main issues are:

- the type of finite element scheme to be used – Section 4.2
- the degree of mesh refinement (if relevant) – Section 4.3.

For frame element models, mesh refinement may not be an issue.

3.3.2   Model options

In some cases it may be best not to develop just one model but to investigate a number of options, evaluate them and choose the one to be used. This is a standard design technique that can be applied to the model in general or to detailed issues, such as the treatment of connections.

3.4   Validation of the analysis model

3.4.1   Validation process

Validation is set out as a discrete event in Table 3.1 but is normally a pervasive activity: features may be validated as the model develops. The validation may not be a major issue for standard situations, but for innovative contexts, and particularly where the risk consequences are high (e.g. long-span roofs), careful validation is essential. The validation work should be recorded as a validation report in the modelling review (Section 3.7).

The process involves making a list of the assumptions for the model and comparing these with the real behaviour of the system. The question to be asked is: is the model capable of satisfying the requirements? Information about the validity

Table 3.4 Validation outcomes

Outcome statement	Context
Criterion satisfied (CS)	A check against a stated criterion was positive, e.g. the span-to-depth ratio of a beam was greater than the minimum
Conventional assumption (CA)	The assumption is standard practice for the type of system being modelled, e.g. neglecting the finite depths of members in a steel frame analysis
Later stage requirement (LSR)	The modelling issue will be satisfied by later actions, e.g. by designing the system to a code of practice – it is important to make sure that such requirements are implemented at the later stages
See sensitivity analysis (SA)	Acceptance is based on information in a sensitivity analysis
To be resolved (TBR)	The final validation decision needs to await further research or use of the model

of the model may be gained from:

- existing records of material testing
- existing records of testing and performance (including failures) of similar systems
- innovative situations, newly commissioned testing work
- sensitivity analysis (Section 2.4.4).

The ideal situation is to define a clear criterion for model acceptance, e.g. using a limiting span-to-depth ratio for a bending member. However, it is not always possible to do this and a range of acceptance outcomes may be used. Table 3.4 gives typical outcomes that may result from a validation of an analysis model.

The validation report can be based on the model development, acceptance criteria and model assurance categories, as set out in Table 3.1 – as in the example on Table 12.2.

3.4.2   Validating the conceptual model
Chapters 4 to 11 give 'validation information' for a range of modelling issues. This is presented in the form of lists of items about validation, but no claim for completeness of these lists is made. These and other issues that may be relevant in the context should be considered.

3.4.3   Validating the computational model
The computational model will solve the conceptual model with some degree of error. The question to be asked is: is the level of error in the computational model acceptable? Two main sources of error in computational models are truncation error (Section 3.5.3) and discretisation error.

Discretisation error is that generated when, after the conceptual model has been defined, the system is divided into finite elements. This creates error from two sources.

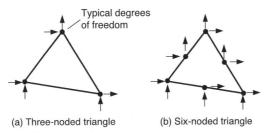

Figure 3.2 *Refinement of plane stress triangular elements.*

- *Element behaviour error* – error due to the assumptions made for the behaviour of the element. For example, a three-noded plane stress triangular element (Fig. 3.2(a)) has constant stress over its area – the simplest possible assumption. This element has a total of six degrees of freedom, but if mid-side nodes are added (Fig. 3.2(b)) the element has a total of 12 degrees of freedom. Therefore, 'higher order', and hence more accurate, functions are used to define its behaviour.
- *Mesh density error* – the mesh density is the number of elements specified per unit area, and it is normally the case that as the mesh density is increased, the exact solution for the conceptual model is approached.

Use of higher order elements and mesh refinement reduce discretisation error. This is a problem of convergence – see Section 4.3.2. Convergence analysis (e.g. see case study, Section 4.4), experience, advice and reference to published information (e.g. from NAFEMS) may be needed in assessing discretisation error.

3.5   The solution process

3.5.1   Selecting software

All software, no matter how well tested, is likely to contain errors. This is a consequence of the fact that it is impractical to test a non-trivial program completely: the number of test cases needed is so vast that they would take forever to run. Testing can help to identify errors, but inevitably bugs are likely to remain.

Practical advice about how end-users can test engineering software can be found in AGS (1994). This document has broad application to software outside of geotechnical engineering. Acceptance criteria could be based on benchmarks published by NAFEMS (1990).

3.5.2   Software validation and verification

The validation question for the software is: is the software capable of solving the conceptual model? This question can be answered by comparing the specification for the software with the conceptual model.

The verification question for the software is: has the software been subject to adequate quality-assurance procedures? For commercial software, answers to this question may be difficult to obtain. Much depends on the reputation of the supplier. When using standard structural analysis packages, software verification is not normally required, although occasional errors do show up in even the best packages.

3.5.3   Truncation error, ill-conditioning

In computer arithmetic, half of the significant figures generated in a multiplication operation have to be discarded. The error in the calculation due to this is called *truncation error*. When a system of equations is susceptible to truncation error it is described as *ill-conditioned*.

This type of error is not common in structural analysis because finite element solutions normally use double-precision arithmetic, which gives 12 or 13 significant decimal digits in each number. However, it can occur, particularly when parts of the structure have widely different stiffnesses, e.g. for a stiff structure on a flexible support (but not for a flexible structure on a stiff support). Slender cantilevers tend to be susceptible to ill-conditioning.

Tests for ill-conditioning

- A check on the equilibrium of the results for a model is, in effect, a check on the residuals of the solution. If the equilibrium is not to an accuracy that is close to that used in the solution (i.e. to 12 significant figures) then, provided that the data is correct, ill-conditioning is likely. Surprisingly, the reverse is not true: satisfactory equilibrium checks do not guarantee absence of ill-conditioning problems.
- The software may perform special calculations to check for ill-conditioning. Details of such checks need to be gained from the software documentation.

3.6   Verifying the results

3.6.1   Acceptance criteria for results

The acceptance criteria for results can be based on risk principles, i.e. on consideration of the combination of likelihood of modelling errors and the consequence of a resulting structural failure – see Table 12.8. The amount of resource to be applied in the checking process will depend on the perceived level of risk. For high-risk situations it may be necessary to commission a complete re-analysis, using different personnel to set up the data and a different software system.

3.6.2   Verification process

The questions to be asked in a results verification are: do the results correspond to what is expected from the model?, or have any errors been made in implementing the solution process?

Whereas validation may be less necessary for standard situations, results verification can never be omitted from the process. There is always significant potential for data errors.

Sources of error

Sources of error in the results include the following:

- *Data errors* – this is the most likely source of error; strategies for ensuring data accuracy need to be adopted.
- *Software errors* – see Section 3.5.2.

- *Hardware errors* – it is sometimes asserted that if there is a hardware error then it is unlikely that sensible results will be produced – this is incorrect. There have been situations where a processor had faults in double-length arithmetic which resulted in output that looked reasonable but was incorrect. However, such situations are now very rare.
- *Truncation error* – see Section 3.5.3.

Checklist
In the results verification process the following checks may be carried out.

- *Error warnings* – check for error warnings in the output.
- *Data check*.
- *Overall equilibrium check* – if the sum of the applied loads on a structure in a given direction is P_a and the sum of the support reactions in that direction is P_s then

$$P_a - P_s = R \tag{3.1}$$

- R should be 0.0, but will normally be a small number, typically of the order of 10^{-12} to 10^{-13}. R is a *residual* of the solution. Reasons for it not being close to zero include:
  - there is ill-conditioning
  - if P_a is calculated using the loading that is expected to be applied (rather than that actually applied as obtained from the software output) and R is not small, then it is likely that there is an error in the loading data.
- *Restraints* – check that the support restraints have been correctly applied by looking at the deformations at the restrained nodes.
- *Symmetry* – if the structure is symmetric then it is worthwhile to apply a symmetric loading case and check the corresponding deformations (or force actions) in relation to a pair of symmetric degrees of freedom. The difference between them should be close to zero, in a similar way to the overall equilibrium check (i.e. use 12-figure accuracy if it is available). If the difference is not small (typically in the last three of the 12 significant digits) then there could be ill-conditioning, but the most likely possibility is that there is an error in the data. For the check to be effective, the symmetry in the definition of the model needs to be precise.
- *Qualitative check* – carry out a qualitative analysis of the results by looking at the deflected shape and the distribution of element forces and stresses. Do these conform with what is expected?
- *Quantitative check* – create a checking model (Section 3.6.3).

3.6.3   Checking models
A checking model is a simplified, more approximate, version of the main model, but which has adequate accuracy for checking purposes (MacLeod 1990). The results from the checking model should show a correlation with the main model results that is consistent with the degree of approximation made for the checking model. The checking model should be used, if practicable, to check both deformations and element forces and stresses.

The checking model can be a simplified model of the system suitable for 'back of an envelope' calculations or it could require a computer solution. If the checking model is a simplified version of the main model, the results may not correspond closely. It is important to be able to justify the degree of difference between the results from the two models. If this cannot be done, then the checking model may have limited value.

Checking models for non-linear analysis can be difficult to formulate. Running a preliminary elastic analysis and using a checking model might be useful for checking the model in general.

Outcomes from using a checking model

Possibilities when comparing the results from a main model and a checking model are as follows.

1. The two results are satisfactorily close. This can mean that:
    (*a*) both results are essentially correct, or
    (*b*) the results appear to correlate, but are in fact in error by similar amounts.
    Outcome (*a*) is the desired situation, but (*b*) can occur and is not easy to identify.
2. The two results are significantly different. In this case work is required to establish the reasons for the difference: the main model may have errors; the checking model may be conceptually wrong or may have errors in the calculation; both models could be in error.
3. The two results are similar, with the main model correct but with the checking model involving compensating but erroneous assumptions or calculations.

In this list, categories 1(*b*) and 3 are *false correlations*. Such situations are remark-ably common. In category 3 the right conclusions may be drawn for the wrong reasons. This may not result in fault, but it is nevertheless unsatisfactory. In an examination for 40 students some years ago, 30% of the correlations between a checking model and an element model were found to be in category 3. People tend to take an optimistic view and when they find that results are close they are quick to accept this as proof of accuracy. A single apparently favourable correlation does not provide a full verification. It is necessary to treat all results with suspicion and not jump to conclusions.

Bracketing the element model results

It is very useful to have an idea of the likely sign of the difference between the main model and the checking model. For example, if the assumptions made for the checking model provide extra stiffness, then one would expect the deflections from it to be basically less than those from the main model. It can be useful to make assumptions for the checking model that are alternately stiffer and more flexible than in the main model. This can then allow a bracketing of the main model's results between upper and lower limits.

Sources for checking models

Descriptions of checking models are given in Sections 5.10.4, 5.11.3, 6.2.4 and 11.6. Examples of the use of checking models are given in Sections 12.1.6 and 12.2.5.

Solutions for checking models may be found in publications such as Young and Budynas (2001).

3.6.4 Checking loadcase

A useful strategy is to use a checking loadcase, where a simplified load (preferably a single-point load, which tends to affect the whole structure) is applied for checking purposes only. Analysis of the results, including the use of a checking model if practicable, of such a loadcase would be carried out before production loadcases are run. However, despite early favourable indications of accuracy, a constant watch for errors must be maintained. Changes made to data (change of member properties, additional loadcases, etc.) have potential for introducing fresh errors.

3.7 The modelling review

3.7.1 Sensitivity analysis

A sensitivity analysis (Section 2.4.4) is an excellent context for:

- helping to understand the behaviour of the system
- providing validation information
- providing information for results verification (e.g. by helping to explain the differences between an element model and a checking model).

While commercial pressure tends to mitigate against doing sensitivity analysis, the value of such exercises should not be underestimated.

3.7.2 Overall acceptance of the results

Before the results from a modelling exercise are passed on to the next stage of the design process it is important that all the results of the modelling activities are considered and used to make the final decision as to whether or not the results should be accepted.

3.7.3 The modelling review document

A *modelling review document* is a collection of information about the modelling activity. The extent of the information to be recorded will depend on the degree to which the analysis is non-standard, the importance of the model and on a risk assessment. The modelling review document for a non-standard analysis of a nuclear facility is likely to be extensive whereas that for a frame in a conventional building would be quite short. Reviews should not contain more information than is necessary.

All of the activities involved in the modelling process may have corresponding items in the review. A main criterion for deciding on the contents of the modelling review is to answer the question: could the results and the rationale for accepting them be checked from the modelling review and data sources without consultation with those who were involved in the modelling exercise?

3.8   Case studies

3.8.1   The Tay Bridge disaster

On 28 December 1899, on a very stormy night, the 13 navigation spans of the Tay Bridge collapsed while a train was traversing them (Fig. 3.3). Seventy-five people died in the disaster, making it possibly the most severe recorded structural failure event in the UK in terms of loss of life. While the collapse occurred long before the computer era, it is worth noting that it was mainly due to a modelling error. The designer, Thomas Bouch, used only $10\,lb/ft^2$ for the design wind pressure when a much larger value should have been used.

At that time there was very little information about the effect of wind loading on structures, and Bouch had to make his own decision about what loading to use. He took advice but used the lowest value suggested to him – lower than he had used for previous designs. He was under considerable pressure from his clients to get the job done quickly and to keep the cost down. It may be that this was a factor in using a low value for wind loading.

Due to modern information on wind loading, the level of risk in design for wind on conventional structures is low, but the commercial pressures on designers has by no means declined. The most important lesson from the Tay Bridge disaster (Martin and MacLeod 1995) for modern times is that one must not allow commercial pressures to erode good engineering practice. There is still the potential for similar disasters. Using the modelling process described in this chapter may involve more engineering time than the client would like, but if it helps to reduce risk, it will be worthwhile.

Figure 3.3   The Tay Bridge structure after the collapse.
Photograph by permission of Dundee City Library.

3.8.2   The Hartford Civic Center roof collapse

The roof over the Hartford Civic Center (Connecticut, USA) was a large-span (110 × 91 m) triangulated steel space deck. On the night of 18 January 1978 it collapsed under snow loading (Levy and Salvadori 1992). Fortunately, the collapse occurred during the night and no one was injured. The 3D space-truss structure had been analysed by computer, but two factors had not been considered:

- Second-order geometry effects (Section 10.2) were not included in the model but were not negligible.
- The axes of the members of the space truss did not intersect at a single point at the connections. The resulting eccentricities caused extra moments in the members (Section 5.6.5).

This collapse was therefore due to faults in the validation of the conceptual model.

3.8.3   The Sleipner platform collapse

The Sleipner offshore platform was under construction in a fjord in Norway in 1990. It was a very large structure – the size of a football field (Fig. 3.4(a)). It had main cells and smaller connecting triangular cells, the 'tricells', as shown in Fig. 3.4(b). The construction was nearing completion, with a tricell full of water with the adjacent drilling cell (D3) partially filled with air. This gave a maximum water-pressure head on the tricell wall of 67 m. Under this loading condition the wall of the tricell collapsed, causing the system to founder. There were no fatalities but the estimated economic cost of the failure was US $700,000.

A main fault in the analysis of the system was that 3D (volume) elements were used to model the complete structure (Foeroyvik 1991). Figure 3.4(b) shows the mesh for a tricell section. There was no refinement of mesh within the depth of the tricell wall and therefore the simulation of bending was probably rather crude. A shell element model for the walls of the tricell might have given better predictions of shear force. At the junctions of the tricell walls, the element

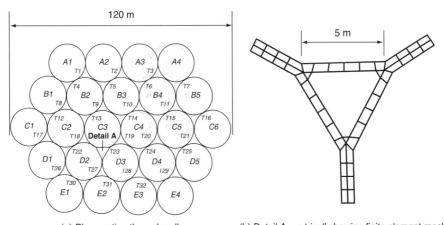

(a) Plan section through cells   (b) Detail A – a tricell showing finite element mesh

Figure 3.4   The Sleipner platform.

shapes were distorted (i.e. the angle between the sides of the elements was not close to 90° – as shown in Fig. 3.4(b). This (according to Foeroyvik 1991) resulted in a 45% underestimate in the predicted shear force in the wall.

There were also problems with the detailing of the reinforcement for the tricells walls, but a major reason for the collapse was faults in the validation of the computational model. Use of a modelling process as discussed in this chapter could have avoided the problem. In particular, a simple checking model (Section 6.3) would have highlighted the problem, as now demonstrated.

Checking model for shear stress in tricell wall of the Sleipner platform
Basis of the checking model
A strip of 1.0 m cross-section width ($b = 1000$ mm) of the tricell wall is assumed to span horizontally.

Operating conditions
Horizontal span of tricell wall $= 4.5$ m; effective depth of concrete in wall, $d = 500$ mm; pressure head, $h = 67.0$ m.

Model calculations
Pressure at 67.0 m depth, p

$$p = \rho g h = 1000 \times 9.81 \times 67.0/1000 = 657 \, \text{kN/m}^2$$

where ρ is the water density.
Load on 1.0 m strip at 67 m depth

$$W = p(4.5 \times 1.0) = 657 \times 4.5 \times 1.0 \times 1000 = 2960 \, \text{kN}$$

Maximum shear force, V

$$V = W/2 = 2960/2 = 1480 \, \text{kN}$$

Shear stress in concrete

$$v_c = V/(bd) = 1480 \times 1000/(51000 \times 500) = 2.96 \, \text{N/mm}^2$$

Maximum design shear stress (for unreinforced section) (BS8110)

$$v_c = 0.91 \, \text{N/mm}^2$$

Outcome
Serious underestimate of shear stress.

4 Modelling with finite elements

4.1 Introduction

This chapter discusses issues relevant to the modelling of structures using finite elements. The term 'finite element' was coined in relation to the analysis of aircraft structures (Turner *et al.* 1956) and the first published finite element stiffness matrix was by Argyris (1954). The conventional approach of dividing a frame structure into engineering beam elements was extended to cover plates in plane stress and biaxial bending. The continuous plates are *discretised* into elements which in themselves only give very rough approximations to the behaviour of the conceptual model but give good accuracy if a sufficient number are used. Since engineering beam elements (Section 5.5.2) do not need such mesh refinement the finite element concept added the dimension of convergence (Section 4.3.2) to element modelling. The term 'finite element method' is now used to cover all element types, whether or not convergence is an issue.

From its introduction in aircraft analysis, the finite element method developed rapidly to encompass a wide range of element types and applications. It became apparent that the method was not only capable of solving structural models but could also be used for field problems, such as heat flow and seepage (Zienciewicz and Cheung 1965). Then it was discovered that it could be used to solve any 'domain' problem that can be expressed in the form of a differential equation with boundary conditions. It now has wide-ranging applications.

The finite element method is a type of computational model (Section 3.3.1).

4.2 Elements

In this section, basic information about the normal range of element types is given.

4.2.1 Constitutive relationships

A *differential element* of the material is defined in relation to dimensions that are differential quantities. For example, Fig. 4.1 shows a differential element for plane stress.

The *constitutive relationship* is the mathematical definition of the behaviour of a differential element of the material. In structural analysis the constitutive relationships tend to have two main sets of assumptions, namely:

- a definition of material behaviour – for example, linear elastic behaviour (Section 7.2) or plasticity (Section 7.3.1)
- assumptions with regard to the stress distribution within the differential element – for example, for the plane stress condition, stresses are defined in

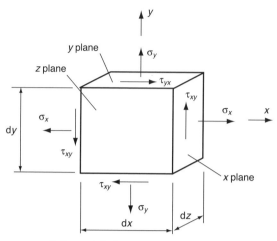

Figure 4.1 Plane stress differential element.

a plane with zero stress at right angles to the plane; for bending, the stresses vary linearly within the depth of the element (see Fig. 5.3).

The constitutive relationship for bending, for example, is a scalar quantity – see equation (5.1). For 2D and 3D elements the constitutive relationship is defined by a square matrix. For example, the constitutive relationship for plane stress is

$$\begin{Bmatrix} \sigma_x \\ \sigma_y \\ \tau_{xy} \end{Bmatrix} = \frac{E}{1-\nu^2} \begin{bmatrix} 1 & \nu & 0 \\ \nu & 1 & 0 \\ 0 & 0 & \dfrac{1-\nu}{2} \end{bmatrix} \begin{Bmatrix} \varepsilon_x \\ \varepsilon_y \\ \gamma_{xy} \end{Bmatrix} \tag{4.1}$$

Where τ, σ_x and σ_y are the direct stresses; τ_{xy} is the shear stress; ε_x and ε_y are the direct strains; γ_{xy} is the shear strain; E is Young's modulus and ν is Poisson's ratio – see Fig. 4.1.

4.2.2 Line elements
Finite element packages normally provide a range of line elements, i.e. elements whose properties are defined along a line. Modelling with line elements is discussed in Chapter 5.

4.2.3 Surface elements
Surface elements have their properties defined over a surface, and include plane stress elements, plane strain elements, plate bending elements and shell elements.

Plane stress elements
The set of stresses used to characterise plane stress is shown in Fig. 4.1. There is no stress and no restraint to movement in the z direction. These are also known as *membrane* stresses.

The case study in Section 4.4 illustrates the use of plane stress elements.

Validation information for isotropic plane stress

Isotropic plane stress is based on the following conditions.

- 'Isotropic' implies that the material has the same properties in all directions.
- The material is homogeneous, i.e. it is uniform.
- The material is continuous, i.e. does not have discontinuities which affect the behaviour.
- The material exhibits linear elastic behaviour (Section 7.2).
- There is no applied stress at right angles to the plane, i.e. $\sigma_z = 0$.
- There is no restraint to strain in the z direction (in the direction of the thickness), i.e. ε_z is unrestrained.
- Values of E and ν are established from tensile test measurements.
- The stresses are uniform across the thickness of the material, i.e. in the z direction.

Validation information for orthotropic plane stress

The validation conditions for isotropic plane stress are relevant to orthotropic plane stress, except that there is a requirement that the material does exhibit properties that are different at right angles.

Plane strain elements

Plane strain elements are used to model a slice of a long structure, such as a dam or an embankment (Fig. 4.2). The condition is close to that of plane stress except that there is no strain but there is stress in the z direction.

Validation information for plane strain

- For plane strain to be valid, the structure must be sufficiently long and uniform such that a slice cut at right angles to the longitudinal axis can be treated as being fully constrained in the direction of the longitudinal axis.
- The validation issues outlined for plane stress also apply to plane strain, with the extra assumption that the strain at right angles to the plane is zero.

Plate bending elements

See Section 6.2.

Shell elements

Shell elements are used to model curved surfaces and flat plates where both in-plane and out-of-plane actions need to be considered – see, for example, the modelling of floor slabs in buildings as described in Section 12.2.3.

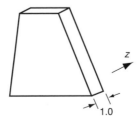

Figure 4.2 Plane strain slice of a dam structure.

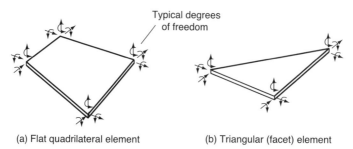

(a) Flat quadrilateral element   (b) Triangular (facet) element

Figure 4.3   Flat shell elements.

Shell elements tend to have six degrees of freedom at each node (Fig. 4.3), taking account of in-plane (membrane) and out-of-plane (bending) actions. They can be flat (Fig. 4.3) or curved. The simplest form of a flat shell element is the combination of a plane stress and a plate bending element.

A triangular flat shell element (sometimes called a *facet element*) can be fitted to any curved surface, but a flat quadrilateral shell element cannot be fitted to a doubly curved shape. For modelling curved shapes it is best to use curved elements.

Validation information for shell elements

Because a shell element can be thought of as a combination of a plane stress element and a plate bending element, the validation information for these (given in Sections 4.2.3 and 6.2.2, respectively) also give guidance for validation of shell elements. Other issues may be relevant and so reference should be made to documentation on the particular element being used.

4.2.4   Volume elements

Volume elements – also called *3D elements* or *brick elements* – i.e. elements defined in three dimensions, tend to be more used in advanced structural analysis because the behaviour of mass structures tends to be non-linear. There are some elastic applications – for example in the modelling of soils (Section 8.3.4).

Figure 4.4 shows typical volume elements, which normally have three degrees of freedom per node. The tetrahedron element corresponds to the triangular and the hexahedron corresponds to the quadrilateral surface elements.

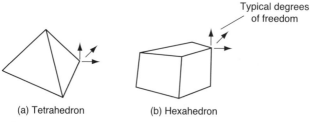

(a) Tetrahedron   (b) Hexahedron

Figure 4.4   Volume elements.

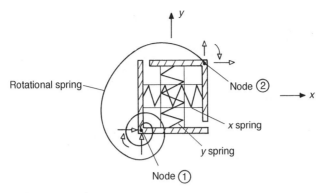

Figure 4.5 In-plane joint element.

4.2.5 Joint elements

A joint element normally comprises a set of springs which connect two nodes or two freedoms. Applications include the modelling of plastic hinges (Section 5.14) and semi-rigid connections (Section 5.6.3). Up to three translational springs and three rotational springs can be included, and various stiffness properties are used.

Figure 4.5 shows a joint element for in-plane conditions. Both nodes would normally be in the same position, but this is not a necessary condition.

Figure 4.6(a) shows a linear elastic spring relationship. This is used, for example, to model partial foundation fixity conditions. Figure 4.6(b) shows elasto-plastic behaviour for a rotational spring. Such a characteristic is used to model plastic hinges in frame structures. Another useful model is the contact spring, which can, for example, take compression but not tension. Figure 4.6(c) shows a more general form of contact spring relationship, which has a low (or zero) stiffness in one direction and a gap.

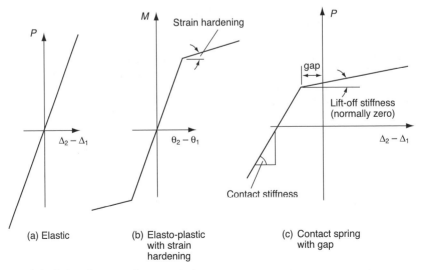

Figure 4.6 Joint element characteristics.

4.2.6 Basic principles for the derivation of finite element stiffness matrices

This section describes some issues that relate to the formulation of element stiffness matrices without digging into the mathematics.

The basic idea is that the assumptions made about the behaviour of an element may be quite approximate in relation to the real behaviour, but by using a large number of elements the correct answer is approached. An analogy for this is the drawing of a series of straight lines to produce a curved line. The greater the number of lines, the more the curve will appear to be smooth.

Basic steps in the derivation of a finite element stiffness matrix include the following points:

- *Define the geometry, the nodes and the degrees of freedom.* Finite elements are defined by shape, nodes and degrees of freedom. Element shapes are discussed in Sections 4.2.2 to 4.2.5. A force applied to a structure and its corresponding deformation are vector quantities, having properties of position, direction and magnitude. The *degree of freedom* is the position and direction that these vectors share. Degrees of freedom normally correspond to translations or rotations, giving a maximum of six degrees of freedom at a node in three dimensions – see Fig. 4.3. Freedoms can correspond to other displacement parameters, but this is not common. It is common to define only translational freedoms at a point, giving two degrees of freedom per node for an in-plane element (Fig. 4.7) or three degrees of freedom per node in three dimensions (Fig. 4.4). The choice of freedoms is somewhat arbitrary, but there are some fundamental requirements – for example, they need to allow rigid body displacements. As an example, Fig. 4.7 shows the degrees of freedom for some plane stress elements. A *node* is a point on the structure at which degrees of freedom are defined.
- *Define the constitutive relationship.* See Section 4.2.1.
- *Define the state of deformation or stress over the area of the element.* It is common to make assumptions about the deformations. For example, for the three-noded plane stress triangle shown in Fig. 4.8 the following displacement functions are used

$$u = \alpha_1 + \alpha_2 x + \alpha_3 y, \qquad v = \alpha_4 + \alpha_5 x + \alpha_6 y \qquad (4.2)$$

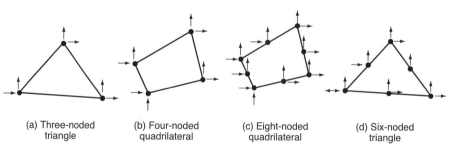

(a) Three-noded   (b) Four-noded   (c) Eight-noded   (d) Six-noded
   triangle   quadrilateral   quadrilateral   triangle

Figure 4.7 Plane stress elements, showing degrees of freedom.

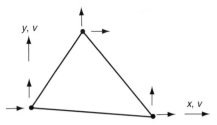

Figure 4.8 Triangular in-plane element.

where u and v are the displacements in the x and y directions and α_i is a constant coefficient.

The strain in the x direction, ε_x, is then $\partial u/\partial x = \alpha_2$, i.e. is constant in this case. The other strain components are similarly constant, and therefore these displacement functions imply that the strain and hence the stress is constant over the area of the element. This is the simplest possible assumption for an element of this type. Some elements have assumptions made about stresses, and in some cases both stress and displacement assumptions are made (hybrid elements).

- *Integration to formulate the stiffness matrix.* This can be done explicitly in some simple cases, but normally numerical integration is used. This is commonly done using *Gaussian quadrature*, where the functions are evaluated at special positions on the element – the *Gauss points*. It is useful to know a little about this process because the output stresses for elements are often given at the Gauss points. For example, for four-point integration of a rectangle, the Gauss points are as shown in Fig. 4.9.
- *Back-substitute to get the element stresses.* Once the equation solver has established the nodal deformations, these are used to calculate the element stresses. For numerically integrated elements the best estimates of stress tend to be at the Gauss points, but the normal requirement is for the average stresses at the nodes. An extrapolation process is therefore required: the stresses at the Gauss points are extrapolated to estimate the stresses at the nodes. This introduces a further degree of approximation to the process.

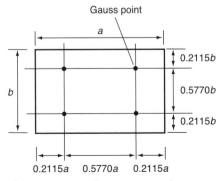

Figure 4.9 Positions of Gauss points for a rectangular area.

It should be noted that numerical integration tends to be favourable to accuracy, whereas numerical differentiation is unfavourable to accuracy. Stresses in finite element modelling are calculated by a differentiation process on the displacements, and hence there is a tendency for predictions of displacement from finite element models to be more accurate than the prediction of stresses.

4.3   Mesh refinement

4.3.1   Discretisation error

The difference between the results from a finite element model (as a computational model) and the 'exact' solution to the conceptual model (Section 3.3.1) is due to *discretisation error* (Section 3.4.3), which is a function of the assumptions made in defining the element stiffness matrix and the mesh density (the number of elements per unit area). The effect of discretisation error is investigated in Section 4.4.

4.3.2   Convergence

Convergence implies that as the mesh density is increased the results will converge towards the exact solution. The rate of convergence depends on the type of element and the number of elements in the mesh. Section 4.4 presents a convergence analysis for a plane stress situation.

Conditions for convergence

The main requirements for an element to converge to the exact solution are that it should allow rigid body movements and allow constant strain states. It is argued that if constant strain is admissible, then by taking smaller and smaller elements, the set of constant strain values can closely approach the sought result.

  Monotonic convergence is where the results from a finite element mesh approach the exact solution in a smooth asymptotic curve as the mesh density is increased (Fig. 4.10). Figure 4.10 also shows an example of non-monotonic convergence.

The patch test

The *patch test* (Irons and Ahmad 1980) is a simple means of assessing the ability of a mesh of elements to model constant strain conditions. Figure 4.11(a) shows a typical patch test model consisting of a mesh of plane stress elements. The overall

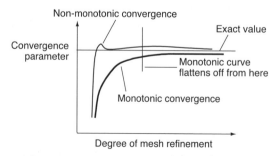

Figure 4.10 Types of convergence.

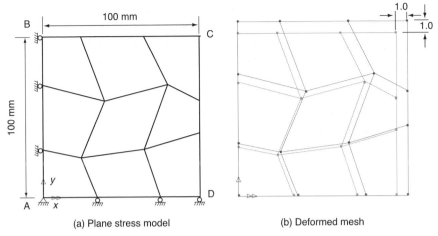

(a) Plane stress model   (b) Deformed mesh

Figure 4.11 Plane stress model for a patch test.

shape is square – 100 × 100 mm – (it could be rectangular) but the element layout is made non regular so as to make the test more effective. A set of fixed deformations is applied to the edges of the model but not to the internal nodes. To simulate this situation, the boundary conditions specified in Table 4.1 were applied.

The conditions of Table 4.1 result in strain applied at the boundaries of 0.01 in both the x and y directions (Fig. 4.11(b)). The requirement of the test is that under the fixed boundary movements the strain will be constant within all elements of the model.

The model of Fig. 4.11(a) was run with linear plane stress elements. The strain in the elements was equal to 0.01 in all cases. With a mesh of quadratic elements the difference between the element strain (ε_e) and the applied strain (0.01) was not greater than 3×10^{-12}. This is at the limits of accuracy of the equation solver and shows that the patch test is satisfied.

When validating an element, it is worthwhile to establish that it can pass the patch test.

4.3.3 Singularities
Re-entrant corner
If a mesh of linear elastic elements is refined at a re-entrant corner, such as for the plate in plane stress shown in Fig. 4.12, the stress at the corner will continue to increase as the mesh density increases. This is because elasticity theory predicts

Table 4.1 Boundary conditions of patch test in Fig. 4.11

Side	Condition
AB	Restrained in the x direction only
AD	Restrained in the y direction only
BC	Fixed deformation of 1.0 mm in the y direction
DC	Fixed deformation of 1.0 mm in the x direction

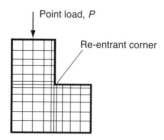

Figure 4.12 Typical situations that cause singularities for a plane stress plate model.

infinite stress at such a corner. This situation is called a *singularity*. The cantilever specimen in the case study in Section 4.4 has a singularity at the corner of the support. The re-entrant corner is well-known in design as a situation of high stress, to be avoided by specifying a radius in the corner or by providing stiffeners.

Point load
Figure 4.12 also shows a point load acting at an edge of the plate. There will also be a singularity for the elastic stresses at the point load. If one considers that there is no specified area of contact for a point load, i.e. the area contact is zero, the contact stress is therefore $\sigma = P/0.0 = \infty$. This is only a problem if local rather than resultant stresses (Section 2.3.1) need to be evaluated. Strategies for getting around the point load singularity problem include:

- define a finite area of contact for the load in the model
- use a non-linear material model.

4.3.4 Benchmark tests
The National Agency for Finite Element Methods and Standards (see NAFEMS 1990) publishes useful benchmark tests. These are situations for which benchmark solutions are available for use to test and compare element performance. When assessing finite element software, it is good policy to make checks against such benchmarks.

4.3.5 Case study – mesh layouts for a cantilever bracket
This case study demonstrates some meshing strategies for a cantilever bracket. Figure 4.13 shows mesh layouts for a bracket with different types of mesh refinement. These layouts are devised to investigate local stresses in the area of the loading.

The layouts are:

- Figure 4.13(a) is a regular mesh without refinement, which would tend not to give good accuracy for local stress.
- Figure 4.13(b) shows a local mesh refinement where triangles are used to diffuse the mesh outwards from the load.
- Figure 4.13(c) does not have the triangular elements of Fig. 4.13(b) and is inadmissible because the node at point X is not connected to the element on its left.

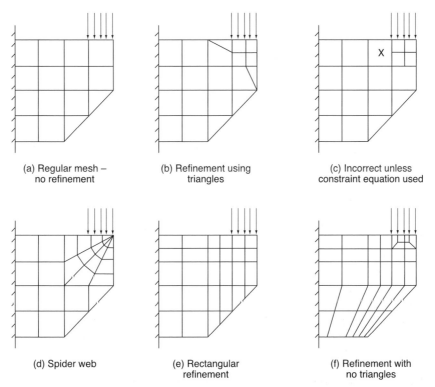

(a) Regular mesh –
    no refinement

(b) Refinement using
    triangles

(c) Incorrect unless
    constraint equation used

(d) Spider web

(e) Rectangular
    refinement

(f) Refinement with
    no triangles

Figure 4.13 Mesh layouts for a cantilever bracket.

- Figure 4.13(d) shows a spider mesh, which is good for keeping the elements as square as practicable, but generating such a mesh might be labour intensive if the program interface does not provide support for such meshes. Note that in the corner under the load, two diagonal lines can be removed to convert triangles to quadrilaterals.
- Figure 4.13(e) shows a refinement using rectangular elements plus a few triangles. This may be the easiest way to set up a refined mesh in this context.
- Figure 4.13(f) shows a mesh refinement strategy with only quadrilateral elements. The mesh under the load is good, but elsewhere there is a lot of distortion in the elements.

4.3.6 Meshing principles

- *Mesh density* – it is best not to use a higher mesh density than is needed in the context. A basic principle is to choose a mesh density at which the convergence curve starts to flatten off (see Fig. 4.10). This is explained in the convergence study in Section 4.4.
- *Lower-order elements* – for example, three-noded triangles and four-noded quadrilaterals (Fig. 4.7). A single quadrilateral element will tend to give significantly better results than would dividing the same area into two triangles. Therefore, triangles should only be used to fill in triangular parts of a surface which cannot be realistically modelled by quadrilaterals. If a line can be

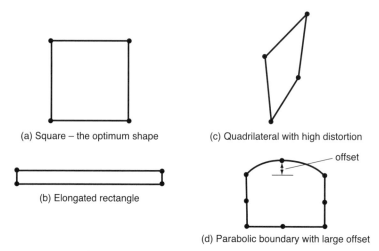

(a) Square – the optimum shape

(b) Elongated rectangle

(c) Quadrilateral with high distortion

(d) Parabolic boundary with large offset

Figure 4.14 Quadrilateral element shapes.

removed to convert two triangles into a quadrilateral then it should be removed (see Fig. 4.13(d)). Constant stress triangular elements can be useful in non-linear analysis since the single stress condition over the surface of the element can be easily tested against criteria (e.g. for plasticity).

- *Higher-order elements* – for example, six-noded triangles (Fig. 4.7(d)) and eight-noded quadrilaterals (Fig. 4.7(c)). These tend to be the most popular types, and although the triangular elements perform quite well it is still best to use quadrilateral elements where practicable.
- *Element shapes* – surface elements tend to give best accuracy when square shapes are used (Fig. 4.14(a)). Elongated shapes and quadrilaterals with angles between the sides which depart significantly from the 90° optimum may give low accuracy (Figs 4.14(b) and (c)). This may be acceptable in areas where the stress gradients are low, but it will be unsatisfactory at stress concentrations. Typical rules are: ratio of side lengths not greater than 2:1; angles between the sides not greater than 135°. For a triangle, the optimum shape is equilateral.
- *Curved boundaries* – some element boundaries can be curved, for example the eight-noded rectangle of Fig. 4.7(c) can have parabolic shaped edges. Accuracy for these elements decreases as the sides' offset from straight increases (Fig. 4.14(d)).
- *Incompatible node at an element edge* – where a node is coincident with the edge of another element which does not have a corresponding edge node, as in Fig. 4.13(c), it is essential to impose a constraint on the deformation of the node and the corresponding position on the element side. If a facility to do this is not available in the software being used then such a meshing arrangement may give poor results.
- *Stress gradients* – where the stress gradient is low then a coarse mesh will be adequate. As the stress gradient increases, a higher density of elements is needed for adequate accuracy. A good strategy is to have a finer mesh in the areas of high stress gradient and a coarser mesh where the stress gradients

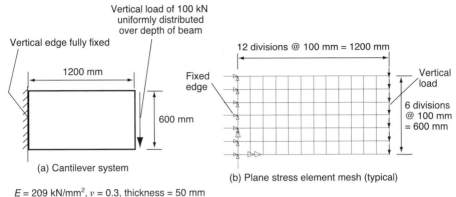

Figure 4.15   Plane stress model for convergence study.

are low. However, for models with a large number of elements it may be best to investigate stress concentrations in special detailed models.

4.4   Case study – convergence analysis of a plane stress cantilever beam model

4.4.1   General

This case study, using a simple plane stress model, demonstrates how the accuracy of a solution is affected by mesh refinement, i.e. discretisation error (Section 4.3.1) in a particular context. The conclusions to be drawn from this study have a degree of general relevance to modelling with other types of element, such as plate bending, shell and volume elements.

4.4.2   The context

Figure 4.15(a) shows a steel plate acting as cantilever beam with transverse end load. This form was chosen to illustrate trends in behaviour rather than to be a realistic example. The span-to-depth ratio of 2:1 could be considered as being unfavourable for accuracy of the engineers' theory of bending (ETB) and a plane stress element model might be used for such a configuration.

Figure 4.15(b) shows a typical 12 × 6 plane stress element mesh for the cantilever beam.

For the main runs, the support conditions were as shown in Fig. 4.15(b), i.e. all nodes on the supported edge are fixed in the *x* and *y* directions.

4.4.3   Elements used in the convergence analysis

Three plane stress element types are included in the convergence analysis. The names used for runs are from the LUSAS (2003)[1] analysis program.

- TPM3 – linear triangular elements with only corner nodes (Fig. 4.7(a)).
- QPM4 – linear quadrilateral elements with only corner nodes (Fig. 4.7(b)).

[1] LUSAS finite element modeller, FEA Ltd.

- QPM8 – quadratic quadrilateral element with corner and mid-side nodes (Fig. 4.7(c)).

All of these elements have two (translational) degrees of freedom per node, as shown in Fig. 4.7

Features of the elements
Features of the behaviour of these elements include the following:

- The elements are all *isoparametric*, based on functions that define both the shape of the element and the displacements. Such elements are *conforming* in that the displacement functions ensure boundary compatibility between the elements. This means that:
  - they give a lower bound to an influence coefficient – i.e. the deformation at, and in the line of, a single-point load on a system of elements will always underestimate the true deformation as compared with an 'exact' solution (Section 2.4.5) to the governing differential equations
  - in general they tend to overestimate stiffness.
- The loading for the system of Fig. 4.15(b) is not a single-point load (it was chosen so as to minimise the effect of local stresses at the free end of the beam) but the arrangement is such that the deformation in the line of the load is very likely to be underestimated by a mesh of any of three element types.
- The TPM3 triangular element has straight sides with a total of six degrees of freedom. It gives constant strain and hence constant stress over the area of the element.
- The QPM4 quadrilateral element has straight sides with a total of eight degrees of freedom. It gives linear variation of strain and hence linear variation of stress over the area of the element.
- The QPM8 element has straight sides (in this context) with a total of 16 degrees of freedom. It gives parabolic variation of strain and hence a parabolic variation of stress over the area of the element.

In this context the *order* of the element is the number of terms in the displacement function used to define the element. For the elements considered here, the number of terms in the displacement function is equal to the number of degrees of freedom assigned to the element. A *higher-order element* (i.e. one which uses higher-order functions and hence a greater number of degrees of freedom) will therefore tend to give more accurate results.

4.4.4   Reference solution
In this case an exact solution is not available. The reference solution used is based on bending theory (Section 5.2) and referred to as the ETB solution. The calculations for end deflection and bending stress (see Fig. 4.16) are as follows.

Parameters

$$E = 209\,\text{kN/mm}^2, \qquad \nu = 0.3,$$
$$A_\text{s} = 5A/6 = 5 \times 6000 \times 50/6 = 25\,000\,\text{mm}^2$$

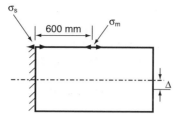

Figure 4.16 Convergence parameters.

where A_s is the shear area and A is the cross-sectional area.

$$I = 50 \times 600^3/12 = 9.0 \times 10^8 \, \text{mm}^4$$

End deflection, Δ

The sum of the bending and shear components

$$\Delta = WL^3/(3EI) + 2WL(1+\nu)/(EA_s)$$

(see Table A4, using $G = E/2/(1+\nu)$)

$$\Delta = \frac{100 \times 1200^3}{3 \times 209 \times (9.0 \times 10^8)} + \frac{2 \times 100 \times 1200 \times (1+0.3)}{209 \times 25\,000}$$

$$= 0.3062 + 0.0597 = 0.3659 \, \text{mm}$$

where W is total end load and L is span.

Bending stress, σ

At support: $\sigma_s = M/Z = \dfrac{100 \times 1200}{(9 \times 10^8)/300} = 0.040 \, \text{kN/mm}^2$

At centre of span: $\sigma_m = M/Z = \dfrac{100 \times 600}{(9 \times 10^8)/300} = 0.020 \, \text{kN/mm}^2$

where M is bending moment and Z is elastic modulus of section.

4.4.5 Convergence parameters

The parameters considered in relation to convergence are:

- Δ^* – the percentage difference between the element model value and the reference model value for the vertical deflection at the centre node of the loaded end – the 'tip deflection', i.e. $\Delta^* = (\Delta_{el} - \Delta_{ETB})/\Delta_{ETB} \times 100$, where Δ_{el} is the value from the element model
- σ_s^* – the percentage difference (defined as for Δ^*) for the horizontal stress at the top of the cantilever at the supported end
- σ_m^* – the percentage difference (defined as for Δ^*) for the horizontal stress at the top of the cantilever midway between the ends.

All results are quoted as being positive.

4.4.6 Meshes

The meshes used for the results given in Fig. 4.17 are shown in Table 4.2. The total number of degrees of freedom, *Dof*, in the mesh is the order of the solution of the simultaneous equations and is therefore a measure of the computing effort needed to achieve a solution.

The calculation of the number of degrees of freedom is demonstrated using the 12 × 6 mesh of Fig. 4.15(b).

- QPM4 – with QPM4 elements there are 12 × 7 nodes, giving 84 active nodes for this mesh (restrained nodes are not counted) with two degrees of freedom per node (Fig. 4.7).

  $$Dof = \text{number of active nodes} \times \text{freedoms per node} = 84 \times 2 = 168$$

- TPM3 – with TPM3 elements, each rectangular part of the mesh is divided into two triangles by a diagonal line. The number of elements is doubled, but the number of degrees of freedom is the same as with QPM4 elements.
- QPM8 – with QPM8 elements there are nodes at all element mid-sides. There are 13 horizontal levels of nodes, giving a total of $24 \times 7 + 12 \times 6 = 240$ nodes. Therefore, $Dof = 240 \times 2 = 480$.

4.4.7 Results

The following observations are based on the results in Fig. 4.17.

Δ^*– Figure 4.17(a)

- For all models the tip deflection converges monotonically towards, and is smaller than, the ETB solution.
- The three-noded triangular element (TPM3) meshes are significantly less accurate than the quadrilateral element meshes. The convergence curves start to flatten off beyond the 12 × 6 mesh.
- The QPM4 curve flattens off beyond the 8 × 4 mesh, giving a result that is close to the ETB value.
- The QPM8 element curve flattens off beyond the 4 × 2 mesh. Even the 2×1 mesh gives a reasonable result.
- The higher the order of element used, the better the accuracy for a given number of degrees of freedom, although the difference is marginal between QPM4 and QPM8.

Table 4.2 Meshes and degrees of freedom (Dof)

Mesh	Dof	
	TPM3, QPM4	QPM8
2 × 1	8	20
4 × 2	24	64
8 × 4	80	224
12 × 6	168	480
24 × 12	624	

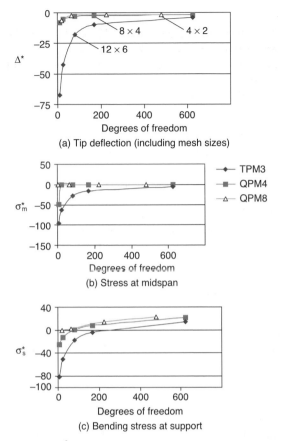

Figure 4.17 Convergence diagrams.

σ_m^* – Fig. 4.17(b)

- The triangular element meshes are significantly less accurate than the quadrilateral meshes but appear to converge monotonically towards the ETB value.
- The QPM4 element gives predictions of σ_m stress which (apart from the 2×1 mesh) are very close to the ETB. The convergence is not monotonic.
- The QPM8 element gives predictions of σ_m stress which are very close to the ETB value for all meshes, but the convergence is not monotonic.

σ_s^* – Fig. 4.17(c)

None of the models give a converging answer. This is because there is a singularity at the fixed corner. Elasticity predicts infinite stress at a singularity, and therefore the finer the mesh, the higher the stress prediction – see Section 4.3.3.

4.4.8 Overview

It is emphasised that the ETB solution is not treated as an exact solution. The accuracy of the element models is assessed not on the basis of how small the differences are from ETB but on how well they compare with the results from the high-density meshes. However, the ETB solution does give very good correlation for the tip deflection and the mid-span (resultant) stress, even with this relatively high span-to-depth ratio.

Of the meshes used for the results in Fig. 4.17, which would be best to adopt in a real modelling situation? In this context the TPM3 triangular elements can be discarded. The 4×2 QPM8 mesh or the 8×4 QPM4 mesh both give good accuracy for deflection. Even the 2×1 QPM8 and 4×2 QPM4 meshes give results for deflection that might be deemed to be acceptable. Apart from the 2×1 QPM4 mesh, it appears that a fairly coarse mesh of rectangular elements gives adequate prediction of resultant bending stress.

Suppose that to analyse the cantilever of Fig. 4.15 the 8×4 QPM4 or 4×2 QPM8 meshes were used. The results for the stress at the support would give very good correlation with ETB and you might treat this as a good correlation, but it would be a false correlation (Section 3.6.3). ETB does not take account of local stresses and therefore cannot pick up the singularity. The plane stress models can identify the singularity, but they cannot approach the correct result because it does not exist at this position. It is common to accept any correlation as a good indicator of validity. Believing that a single correlation validates a model can be dangerous because false correlations are not uncommon. Convergence analysis and sensitivity analysis are important tools in avoiding acceptance of such correlations.

4.5 Constraints

4.5.1 General

A *constraint* is a condition imposed on the deformations of a structure – effectively a compatibility condition. For example, the imposition of continuity between members meeting at a joint in a frame is a constraint. A *restraint* (Section 8.2.1) can be viewed as a constraint to deformations at a support.

Sometimes it is advantageous to impose a rigid body movement on a part of the structure – denoted here as a *rigid constraint*. See, for example, the modelling of composite slabs (Section 6.3.4).

4.5.2 Rigid constraint conditions

Constraints can be incorporated into the model using the following techniques.

- Constraint equations – see Section 4.5.3.
- A rigid link, which is incorporated into the element stiffness relationships.
- A beam element, with high but finite bending stiffness. This is an effective approach but it is important not to make the stiffness too high otherwise there may be numerical conditioning problems in the solution (Section 3.5.3). The bending stiffness of the 'rigid' member, $(EI)_r$, can be defined as

$$(EI)_r = c_r (EI)_{max} \tag{4.3}$$

where c_r is a coefficient in the range 10–100 and $(EI)_{max}$ is the highest EI value of the members which are connected to the rigid member (or the sum of the EI values of the members at the connection).

4.5.3 Constraint equations
Master and slave nodes
A useful concept for constraint equations is that of 'master' and 'slave' nodes. For example, the displacements of a rigid domain (line, area or volume) can be defined by the displacements of a single node – the *master node*. The displacements of the other nodes in the domain – the *slave nodes* – can be related to those of the master node using only the geometry of the domain (i.e. no stiffness relationships are needed). An example of the basic mechanics of the constraint equations for a rigid line is now described.

Rigid lines
In Fig. 4.18, if node 1 is designated as the master node and node 2 is a slave node then the displacements at these nodes will be related by

$$\Delta_{x2} = \Delta_{x1} \tag{4.4}$$

$$\Delta_{y2} = \Delta_{y1} + L\theta_1 \tag{4.5}$$

$$\theta_2 = \theta_1 \tag{4.6}$$

where the displacements are defined in Fig. 4.18.
   These relationships expressed in matrix notation are

$$\begin{Bmatrix} \Delta_{x2} \\ \Delta_{y2} \\ \theta_2 \end{Bmatrix} = \begin{bmatrix} 1 & 0 & 0 \\ 0 & 1 & L \\ 0 & 0 & 1 \end{bmatrix} \begin{Bmatrix} \Delta_{x1} \\ \Delta_{y1} \\ \theta_1 \end{Bmatrix}$$

that is

$$d_{\mathrm{s}} = A d_{\mathrm{m}} \tag{4.7}$$

where d_{m} are the master displacements at node 1, d_{s} are the slave displacements at node 2, and A is the constraint matrix.
   Note that in defining the constraint relationship, the small deformations condition of $\theta \approx \tan \theta$ is used.

Checking constraint conditions
It is important to check that constraint conditions have been properly implemented. A quick check on the constraint equation can be made by checking that the rotations at the slave nodes are the same as those at the master node, e.g.

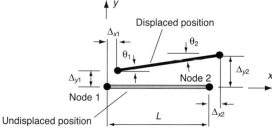

Figure 4.18 Rigid movement of a line.

using equation (4.6). This does not provide a full check: the translational conditions may also need to be checked. This can be done by setting up the relevant constraint equation for a master and a slave node – e.g. equations (4.4) and (4.5) are the translational restraint equations for a horizontal rigid line – and substituting the values of displacements from the results.

If a constraint equation has been used then the conditions should be satisfied to the accuracy of the solution of the stiffness equations (Section 3.5.3). If a high but finite stiffness has been used then the accuracy to which the relationships should be satisfied will depend on the value of c_r used (equation (4.3)).

4.6   Symmetry

4.6.1   General

If a structure has an axis of symmetry then the order of solution can be reduced. In this section, the principles involved in taking account of mirror symmetry are discussed. For more information on the use of symmetry see NAFEMS (1992). Whether or not symmetry is used to reduce the order of solution, developing an understanding of symmetry conditions is worthwhile as it is very useful for checking.

While the need to reduce the size of models is now less important due to the high level of computing power available, there may be circumstances where the use of a symmetric model will be advantageous. I have used a symmetrical 1/8 model of a pyramid structure (using the orthogonal and diagonal axes of symmetry) and a symmetrical 1/4 of a slab structure to advantage.

4.6.2   Mirror symmetry
Conditions for mirror symmetry
In order for mirror symmetry to be satisfied, all geometric and material properties and all loading must be the same at corresponding points on either side of the axis of symmetry.

Two types of loading in relation to mirror symmetry are:

- *symmetric loading*, Fig. 4.19(a) – results in displacements at the axis of symmetry in the line of that axis
- *antisymmetric* (also called *skew-symmetric*) *loading*, Fig. 4.19(b) – results in displacements at the axis of symmetry at right angles to the line of that axis.

Modelling for mirror symmetry
Figure 4.20(a) shows how a symmetric equivalent condition is modelled, and Fig. 4.20(b) shows an antisymmetric equivalent. The cross-sectional properties

    (a) Symmetric loadcases   (b) Antisymmetric loadcases

Figure 4.19 Symmetric and antisymmetric loadcases.

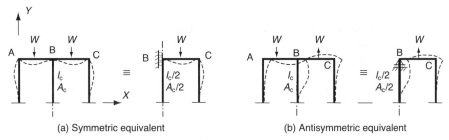

Figure 4.20 Symmetric and antisymmetric equivalent models.

of the member on the axis of symmetry of the symmetrically (or antisymmetrically) equivalent model are half of those for the complete frame, as shown in Figs 4.20(a) and 4.20(b). The restraints at the axis of symmetry are defined as in Table 4.3. Note that the restraint conditions are opposite to each other for the two cases.

The results from a symmetric equivalent frame can be interpreted as follows.

- The *displacements* in the equivalent frame will mirror those of the full frame in a symmetric or antisymmetric way, depending on the loadcase.
- The *internal force actions* will be mirrored as for displacements. The force actions in the member of the equivalent frame which lies on the axis of symmetry should be multiplied by a factor of two to emulate those of the full frame.

Defining restraints for mirror symmetry models

The following is a process (which applies to 2D and 3D situations) for defining restraints to a symmetrical or antisymmetrical part of a structure so as to model only a symmetrical part.

1. Identify the nodes that lie on the axis (axes) of symmetry.
2. For each displacement component at these nodes, specify a restraint if the deformation will be zero under the symmetric (or antisymmetric) loading condition.
3. If such deformations are not zero then they should be unrestrained (unless it is a support).
4. For members which have longitudinal axes on an axis of symmetry, the cross-sectional properties used in the symmetric part should be one half of those in the actual member.

Table 4.3 Restraints for the models of Fig. 4.20

Model	Node	Restraint		
		X	Y	θ_z
Symmetric	B	R	F	R
Antisymmetric	B	F	R	F

Notes: F – free; R – restrained.

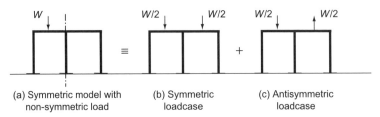

<table>
<tr><td>(a) Symmetric model with
non-symmetric load</td><td>(b) Symmetric
loadcase</td><td>(c) Antisymmetric
loadcase</td></tr>
</table>

Figure 4.21 Treatment of a non-symmetric load on a symmetric structure.

Modelling a non-symmetric loadcase using mirror symmetry

Figure 4.21(a) shows a symmetric frame with a non-symmetric load W. By summing a symmetric loadcase (Fig. 4.21(b)) and an antisymmetric loadcase (Fig. 4.21(c)) the loading on the right-hand beam cancels out and the loading on the left hand beam is W. Symmetric equivalents can be used in both cases; the order of solution is halved at the expense of two separate runs with different loading and boundary conditions.

4.6.3   Symmetry checking

If the structure has axes of symmetry but is modelled in full, a symmetry check can be helpful in verifying the input data. A special symmetric (or antisymmetric) checking loadcase (Section 3.6.4) can be used. First of all, check that the displacements at the axis of symmetry correspond to the symmetry condition. Then, compare displacements or internal actions at a symmetrical position. It is normally only necessary to check one item, but the correspondence between the symmetric values must be satisfied to the accuracy of the solution of the stiffness equations – see the example in Section 12.1.6. If the values do not match in this way then the symmetry condition is violated. Such lack of symmetry may not be important but the reason for it should be identified. It could be that the geometry on either side is not the same, which might not be important but it could be due to, for example, columns on one side of the axis of symmetry being oriented differently to those on the other side. Such errors need to be fixed. Lack of symmetry in the results when they should be symmetric can also be due to truncation error (Section 3.5.3).

5 Skeletal frames – modelling with line elements

5.1 Introduction

The members of a skeletal frame (Fig. 5.1) have length-to-width ratios that are sufficiently high to define their structural behaviour by elements whose properties are defined along a line – i.e. *line elements*. In this chapter, general issues in relation to line elements and the modelling of skeletal frames are discussed.

The two main types of line element normally defined in software are *bar elements* and *beam elements*. The bar element is an axial force element which forms part of a beam element.

Figure 5.1 Skeletal frame. Photograph by permission of Jacobs Babtie.

In order to understand the behaviour of beam elements it is necessary to look at the components – bending, axial effects and torsion. The information given on these topics in Sections 5.2, 5.3 and 5.4 is about modelling issues; only a limited amount of information on the basic mechanics is provided.

5.1.1   Members and elements
In this context, the following definitions are used.

- *Member* is a physical part of a structure, such as a column or a beam.
- *Element* is a mathematical representation of a member normally using line elements for skeletal frames. The analysis model of a member may comprise several elements.

5.2   Bending

5.2.1   Background
In 1826, Louis Navier, a professor at the École des Pont et Chaussées in Paris, published his *Leçons* on the use of mechanics in the design of buildings and machines. Straub (1964) wrote that this made him the "creator of that branch of mechanics which we call structural analysis". This was a major event in the Industrial Revolution. Prior to that time, the design of a beam would be based on heuristic rules, for example limiting span-to-depth ratios. By using bending theory as presented in *Leçons*, it became possible to design a beam with greatly reduced risk. Reliability increased, cost decreased, applications burgeoned.

Basic bending theory is often called the 'engineers' theory of bending' (ETB).

5.2.2   Behaviour
Figure 5.2(a) shows the deformed mesh for a plane stress finite element model of a simply supported beam with uniformly distributed loading. Note that:

- the vertical lines tend to remain remain straight in the deformed mesh (basic assuption of bending theory) but are less straight near the supports, where shear deformation is more in evidence (Section 5.2.5)
- point *a* is at a roller joint and moves outwards.

Figure 5.2(b) shows the typical deformed shape of an element in bending. Note that the sides remain straight. This is the main assumption for bending theory, i.e. that plane cross sections remain plane after deformation.

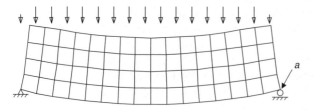

(a) Deformed mesh due to uniformly distributed load   (b) Deformed shape of element

Figure 5.2   Plane stress model of a beam in bending.

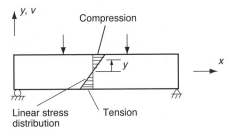

Figure 5.3 Bending stress in a beam.

Figure 5.3 shows the distribution of longitudinal stress in a beam. It tends to be linear in the depth of the beam, with (in this case) compression at the top and tension at the bottom. The level of zero bending stress is at the neutral axis, which coincides with the centroidal axis when there is no resultant axial force on the section.

5.2.3 Basic relationships for bending
Force-deformation
The governing equation for bending about a principal axis of a section composed of a single material is

$$M = EI \frac{\mathrm{d}^2 v}{\mathrm{d}x^2} \tag{5.1}$$

where M is the bending moment, E is Young's modulus (Table A6), I is the second moment of area (Table A1) and v is the deformation in the y (vertical) direction.

Stress
The direct stress in the axial direction, σ, of the beam is given by

$$\sigma = My/I \tag{5.2}$$

where y is the distance from the centroidal axis to the level at which stress is to be defined.

Bending theory can also be applied to composite cross sections, i.e. to sections composed of more than one material (see texts on structural mechanics listed in the bibliography, Chapter 5).

5.2.4 Symmetric and asymmetric bending
Figure 5.4(a) shows a cross section that is symmetric about the y axis but not about the z axis. If it is loaded in the y direction (i.e. in the xy plane) it will deflect in the y direction but not in the z direction – it will be in symmetric bending for this loading. If it is loaded in the z direction (i.e. in the xz plane) it will tend to deflect in the z and in the y directions – it will be in asymmetric bending.

The section of Fig. 5.4(b) does not have an axis of symmetry and therefore would be in asymmetric bending if loaded in the y or the z directions.

If the axis about which the plane of loading is rotated is as in Fig. 5.4(b), a situation will be identified where the load does not cause displacements at right angles to

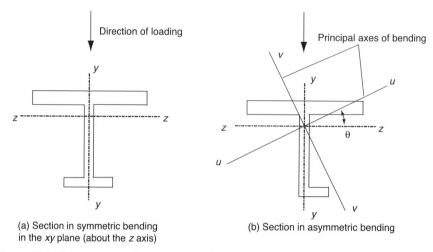

(a) Section in symmetric bending
in the *xy* plane (about the *z* axis)

(b) Section in asymmetric bending

Figure 5.4 Symmetric and asymmetric bending cases.

the plane in which it acts. This axis and the axis at right angles to it are the *principal axes* of the section.

For asymmetric sections it is necessary to identify the principal axes, transform the loading into the principal directions, and then calculate the stresses by reference to these axes.

5.2.5   Shear in bending
Behaviour
Figure 5.5(a) shows the deformed mesh for a plane stress model of a deep cantilever with a uniformly distributed line load acting along line *a–a*. There is some bending action, but the behaviour is dominated by shear deformation.

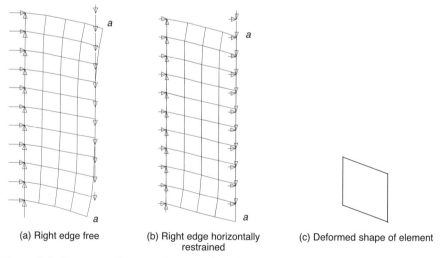

(a) Right edge free

(b) Right edge horizontally
restrained

(c) Deformed shape of element

Figure 5.5 Deep cantilever with vertical uniformly distributed edge load.

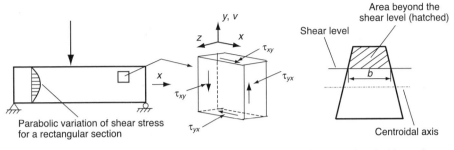

(a) Stress distribution and directions   (b) Cross section of arbitrary shape

Figure 5.6 Shear stresses in a beam.

Figure 5.5(c) shows an element under shear deformation. Note that line *a–a* does not remain straight in Fig. 5.5(a); there is significant warping of the cross section which is not consistent with the plane sections remain plane assumption of bending theory. This type of behaviour is always present in beams, but it tends to be negligible except with low span-to-depth ratios.

Figure 5.5(b) shows the same model with line *a–a* constrained horizontally. This is close to a 'pure shear' situation, although the shear strain must be zero at the top and at the bottom where the *x* direction and *y* direction mesh lines meet at right angles.

Figure 5.6(a) shows the parabolic variation of shear stress across a rectangular section. The shear stress in the web of an I section is a flat parabola, but the distribution is normally assumed to be uniform.

Basic relationships for shear in bending
Figure 5.7 shows a differential element of a material under shear deformation. The shear deformation is defined by the angle dv/dx. (Note that in the more general case the shear is $\partial v/\partial x + \partial u/\partial y$ but only the dv/dx component is considered in bending behaviour.)

Force deformation
For a beam in bending, the shear stiffness relationship is defined as

$$S = K_s\, dv/dx \tag{5.3}$$

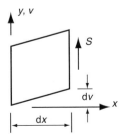

Figure 5.7 Shear deformation.

where S is the shear force, $K_s = A_s G$ is the shear stiffness, A_s is the shear area (see Table A2), G is the shear modulus $= E/(2(1 + \nu))$ (equation (7.6)), dv/dx is the shear deformation (an angle) and v is the displacement in the y direction.

Shear stress
The shear stress, τ, at a section of a beam is

$$\tau = sQ/(Ib) \qquad\qquad (5.4)$$

where Q is the first moment of area of the section beyond the shear level (Fig. 5.6(b)) about the centroidal axis of the section, the *shear level* is the level at which the shear stress is to be calculated (Fig. 5.6(b)), I is the second moment of inertia and b is the width of the section at the shear level (Fig. 5.6(b)).

Note that there is both a vertical and a horizontal component of shear stress. Complementary shear stresses τ_{xy} and τ_{yx} of equal magnitude act on the x and the y planes (Fig. 5.6(a)).

5.2.6   Combined bending and shear
It is normally assumed that for the displacement of a beam, the components due to bending and shear deformation can be added directly. Therefore a particular beam displacement Δ can be calculated using

$$\Delta = \Delta_b + \Delta_s \qquad\qquad (5.5)$$

where Δ_b is the displacement due to bending and Δ_s is the displacement due to shear. The in-series addition of the bending and shear components can be implemented in the bending stiffness matrix for the beam element – sometimes call a 'thick beam' – see Section 5.5.2.

5.2.7   Validation information for the engineers' theory of bending
- *Linear elasticity* – see Section 7.2.
- *Resultant stresses* – the stresses predicted by bending theory are resultant stresses (see Section 2.3.1). Local stress concentrations are not predicted by bending theory.
- *Symmetric or asymmetric bending* – if the plane of loading is not a principal plane of the cross section then the orientations of the principal planes have to be identified and the loading transformed into these planes.
- *Plane sections remain plane* – i.e. no warping of the cross section (warping is the change of shape of the section out of its plane – Fig. 5.2). With pure bending (no shear force and hence no shear deformation), this assumption tends to give very good results for the calculation of resultant stress and displacement due to bending. With shear force, the effect of shear deformation may be significant.
- *Shear deformation* – shear deformation assumes that cross sections do warp (Fig. 5.5(a)). The ETB models for shear deformation and bending deformation therefore have opposing assumptions. Despite this, adding the effect of shear

deformation to the bending deformation (equation (5.5)) tends to give good results for predicting overall deflection, even for low span-to-depth ratios – see Section 4.4.7. Shear deformation tends to affect the prediction of displacement more than it affects the prediction of stress. Factors which affect the ratio of displacement due to bending deformation to that of shear deformation include:

o the span-to-depth ratio – the displacement due to shear deformation becomes more significant as the span-to-depth ratio decreases

o the shape of the cross section – for example, the addition of flanges to a rectangular beam can have a very significant effect on the displacement due to bending but does not significantly affect that due to shear deformation (flanged beams are therefore more susceptible to the effect of shear deformation than rectangular beams)

o the distribution of the shear force in the beam – this tends to have a secondary effect.

Typically, if the span-to-depth ratio is greater than 10 it is very unlikely that shear deformation will need to be considered. The following approximate relationship can be used as a rough measure of the relative effect of bending to shear deformation

$$\Delta_b/\Delta_s = 0.04 A_s L^2 / I \qquad (5.6)$$

where Δ_b and Δ_s are the central deflections due to bending deformation and shear deformation respectively for a simply supported beam with uniformly distributed loading, A_s is the shear area of the beam, L is the span and I is the second moment of area of the cross section. (Equation (5.6) is based on the formulae given in Table A4.)

• *Small deformations are neglected* – in the derivation of equation (5.1) the curvature is approximated by the expression

$$\frac{1}{R} = \frac{d^2 v}{dx^2}$$

rather than

$$\frac{1}{R} = \frac{d^2 y/dx^2}{(1 + (dy/dx)^2)^{3/2}}$$

This assumption is valid provided that the rotation of the beam (dy/dx) is low. The normal limiting deflection-to-span ratio $(1:300)$ results in rotations that are sufficiently small for this assumption to be valid.

• *Neglect second-order effects due to flexural-torsional buckling* – these effects can be very important for open sections, such as I sections. They are not normally included in frame models. Code of practice rules for sizing of members normally take account of these effects. For further information see Trahair (1993).

• *Concrete sections* – it is common when modelling concrete sections to use the gross concrete section, neglecting the effect of reinforcement on the stiffness.

This may significantly overestimate the stiffness of concrete beams with cracked sections. For rules for taking account of cracking see Eurocode 2 (CEN 2004a).

5.3 Axial effects

5.3.1 Behaviour

If the resultant of the applied load on a uniform straight member is in the direction of the longitudinal axis of the element and acting at the centroid of the section then the resultant axial stress in the member will tend to be uniform. This is the basis of the axial component in a beam element.

Figure 5.8 shows a plane stress model under uniaxial loading. Point a is pinned and points b are fixed in the x direction but free in the y direction. A uniformly distributed load acts horizontally at line c–c. Note that:

- the uniform tensile deformation in the x direction corresponds to the uniform applied load
- the compressive movement in the y direction is due to the Poisson's ratio effect (the shape of this model is chosen to illustrate this effect).

Figure 5.9(a) shows a plane stress model with a more realistic length-to-breadth ratio. The left-hand end has the same type of support as for the model in Fig. 5.8, but the load at the right end is a point load in the centre. Note how in Fig. 5.9(a) the cross section warps near the load but deforms uniformly at positions remote from the load. Figure 5.9(b) is a contour plot of horizontal stress with the point load. Note the stress concentrations in the vicinity of the load and the constant stress elsewhere. The axial action in a beam element ignores the local stresses and deformations and only includes the resultant uniform actions (see Section 2.3.1).

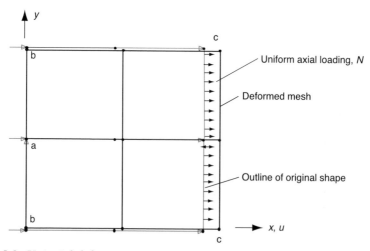

Figure 5.8 Uniaxial deformation.

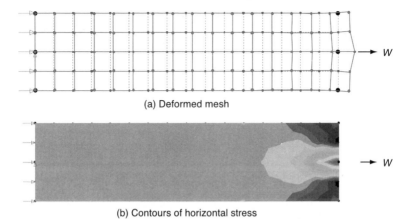

(a) Deformed mesh

(b) Contours of horizontal stress

Figure 5.9   Plane stress model with point loading at the free end.

5.3.2   Basic relationships
Force deformation
The governing differential equation for axial action is (Fig. 5.8)

$$N = EA \, \mathrm{d}u/\mathrm{d}x \tag{5.7}$$

where N is the axial force, E is Young's modulus, A is the cross-sectional area (see Table A1) and u is the deformation in the direction of the longitudinal (x) axis.

Stress
The uniform axial stress σ_x is given by

$$\sigma_x = N/A \tag{5.8}$$

5.3.3   Validation information

- *Linear elasticity* – see Section 7.2.
- *Resultant actions* – only resultant actions are modelled (see Section 2.3.1).
- *Uniform stress over the section* – the resultant of the axial force acts at the centroid of the section area, giving uniform resultant stress in the element. Moment due to eccentricity of the axial load is ignored (part of the bending component of a beam element).
- *First-order analysis* – the axial effects as characterised by equations (5.7) and (5.8) are assumed to be uncoupled from the bending and torsional components.
- *Second-order analysis* (i.e. non-linear geometry) – the effect of the eccentricity of the axial load due to the bent shape of the element can be taken into account (see Chapter 10). A typical acceptance criterion for neglecting the second-order effects in a member is

$$N/N_{\mathrm{cr}} < 0.1 \tag{5.9}$$

where N_{cr} is the Euler buckling load (see Section 10.3.1).

5.4   Torsion

5.4.1   Behaviour

Warping of the cross section

An important issue in torsional deformation is whether or not the cross section warps. *Warping* is deformation of the cross section out of its plane. For example, the cross section of the I section under torsion shown in Fig. 5.12 has warped.

Shear torsion

Figure 5.10 shows a rod with a circular cross section, fixed at the lower end with a torque T at the other end. An element of the material is in a state of pure shear (the principal direct stresses – at 45° to the x axis of the bar – have equal and opposite values). Straight lines parallel to the x axis (longitudinal axis) remain straight after deformation. For example, line b–a on Fig. 5.10 is the undeformed position of a straight line parallel to the x axis at the surface of the rod. After T is applied, the line moves to b–a', which also tends to be straight. With torque applied only at the free end, the shear stress is constant with height and varies linearly in the radial direction from zero at the centre. This is the basis of the theory for bending torsion.

Figure 5.11 shows a 3D finite element model of a bar with a square section under the same conditions of support and loading as that of Fig. 5.10. In this case there is warping of the cross sections and the lines in the x direction do not remain straight.

Bending torsion

Figure 5.12 shows an I section with one end fully fixed and a torque T applied at the other end. If T is modelled as a pair of equal and opposite forces ($= T/D$, where D is the distance between the flanges of the section) acting at the top of the flanges then each flange is loaded by a top point load, as shown in Fig. 5.12. The tops of the flanges will rotate in opposite directions and therefore the cross section deforms out of its original plane – this is *warping* of the cross section.

This is an example of *bending torsion* – also known as *warping torsion* and *non-uniform torsion*. Its presence requires warping restraint (restraint to the bending

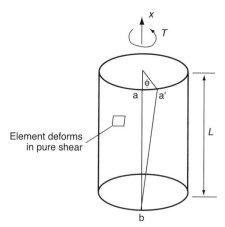

Figure 5.10   Circular rod with end torque.

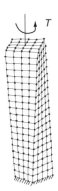

Figure 5.11 Square bar in torsion.

action in the flanges in this case), as is provided by the fixity at the lower end of the I section in Fig. 5.12.

Bending torsion is mainly applicable to members whose cross section is made up of thin-wall, open section plates.

Distortion
With a thin-walled section, the cross section may change shape in its own plane (Fig. 5.13) – this is termed *distortion* (as distinct from warping, which is out of plane). This can be important in some circumstances, for example box sections, but is not allowed for in conventional beam elements.

5.4.2 Basic relationships for shear torsion
This theory is also known as the 'St Venant theory of torsion' or the 'engineer's theory of torsion' (ETT).

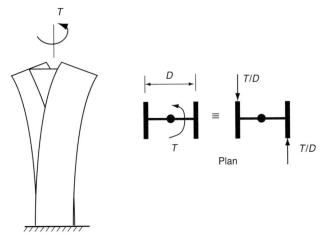

Figure 5.12 I section with end torque.

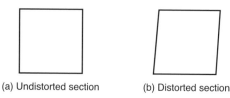

(a) Undistorted section   (b) Distorted section

Figure 5.13 Distortion of cross section.

Force deformation

The governing equation for shear torsion is

$$T = GJ \, d\phi/dx \tag{5.10}$$

where T is the torque, G is the shear modulus ($G = E/2/(1 + \nu)$) (equation (7.6)), J is the shear torsion constant for the cross section (see Table A3) and ϕ is the rotation about the x axis – Fig. 5.10.

Stress

See Table A3.

5.4.3   Basic relationships for bending torsion

Force deformation

The governing equation for bending torsion is

$$dT/dx = EI_\omega \, d^4\phi/dx^4 \tag{5.11}$$

where I_ω is the second moment of sectorial area and ϕ is the rotation of the section.

Stress

The direct stress due to bending torsion is given by

$$\sigma = B\omega/I_\omega \tag{5.12}$$

where B is the applied bimoment and ω is the sectorial co-ordinate.

Bimoment

Vlasov (1961) devised the concept of a *bimoment* to define the internal forces for bending torsion. Figure 5.14 shows two equal and opposite bending moments, M, acting about the same axis at a distance apart, D. The bimoment,

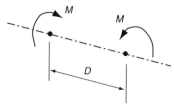

Figure 5.14 Definition of a bimoment.

B, has the value

$$B = MD$$

That is, the bimoment is the product of the value of the bending moments times the distance between the points at which they act. For example, in Fig. 5.12 the bending moment in the flanges at the fixed end is $(T/D)L$. The bimoment at this level is therefore $(T/D)L \times D = TL$.

The concept of a bimoment seems difficult at first but its application is directly analogous to bending theory. Note the correspondence between equation (5.12) – $\sigma = B\omega/I_\omega$ – and the bending equivalent, equation (5.2) – $\sigma = My/I$.

The calculation of sectorial properties is more complex than for other sectional properties but can be automated. For sources of information on bending torsion see the bibliography, Chapter 5.

5.4.4   Combined torsion

The governing equation for combined shear torsion and bending torsion is

$$\frac{dT}{dx} = GJ \frac{d^2\phi}{dx^2} - EI_\omega \frac{d^4\phi}{dx^4} \tag{5.13}$$

The torque, T, is considered to be resisted by two components:

- T_t – torque taken by the shear torsion
- T_ω – the torque taken by the bending torsion.

Torque is therefore

$$T = T_t + T_\omega \tag{5.14}$$

To include bending torsion in an element it is normal to have an extra degree of freedom, which corresponds to a bimoment at each node (Taranath 1998, McGuire *et al.* 2000).

5.4.5   Validation information for torsion

The main assumptions are as follows.

- *Linear elastic behaviour* – see Section 7.2.
- *Resultant actions* – only resultant actions are modelled (see Section 2.3.1).
- *Shear torsion* – axial lines remain straight after deformation. This gives high accuracy for circular rods with high length-to-diameter ratios. Shear torsion theory is generally acceptable for closed sections (e.g. circular and rectangular hollow sections) and for solid square and rectangular sections.
- *Bending torsion* – the accuracy of bending torsion theory will increase as the length-to-width ratios of the plates that make up the section increase. A major difficulty can be the definition of the degree of warping restraint. The main assumptions for bending torsion theory are:
  - the plates which form the cross section deform in bending in their own planes
  - out-of-plane bending of the plates is neglected
  - shear deformation is neglected

- the plates are continuously connected to each other longitudinally
- the layout of the plates of the cross section must be such that a bimoment can be resisted – if the planes of all the plates intersect at a single point (e.g. angle and T sections) then there can be no bimoment.

- *No distortion of the cross section* – distortion is change of shape of the cross section in its own plane (see Fig. 5.13). This may not be negligible for some thin-walled sections.
- *Second-order effects* – interaction between torsion and axial and bending effects of the beam element is neglected. Torsion does interact with bending in lateral torsional buckling – this is not normally taken into account in frame models but tends to be covered by code of practice rules.

With circular cross sections and with no warping restraint there is no bending torsion, but otherwise both shear and bending torsion will be present if there is an applied torque. The normal model for torsion in beam elements is to take account only of shear torsion. This will tend to give acceptable results for closed sections, but it will be significantly inaccurate (not conservative) for open sections. Using only shear torsion for a member with an open section is only acceptable if torsion has a negligible effect on its structural behaviour, i.e. if torsion can effectively be ignored for the member. This may be assessed using a sensitivity analysis (Section 2.4.4).

If an open section has to be designed to take torsion then bending torsion may have to be included. Structural designers mostly avoid this situation by using closed sections for torsion, although cases arise where bending torsion needs to be considered, for example in curved I-section or channel staircase supports. In this situation, an element that includes bending torsion can be used. For further information see the bibliography, Chapter 5.

5.5   Bar elements and beam elements

5.5.1   Bar elements

The main type of bar element is straight with only one axial freedom at each end. Such elements are typically used to model pin connected struts. The conventional bar element, with freedoms in local co-ordinates (as shown in Fig. 5.15), has a single axial freedom at each of the two end nodes.

The freedoms for a bar element, shown in Fig. 5.15(c), shows the freedoms in local co-ordinates. In system co-ordinates, the single axial freedom is transformed to:

- two freedoms in the global directions in a plane
- three freedoms in the global directions in space.

Higher-order bar elements with a mid-length node are sometimes used for special purposes. For validation information for bar elements see Section 5.3.

5.5.2   Engineering beam elements

Figure 5.15 shows the basic engineering beam elements. These elements are straight, with constant properties along their lengths. They have the important

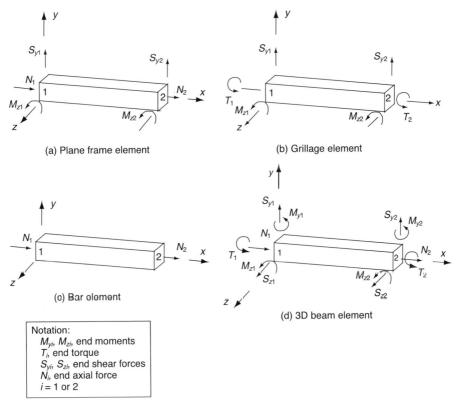

Figure 5.15 Engineering beam elements.

feature that they do not need mesh refinement to improve accuracy. This is because their stiffness matrices represent correct (as distinct from approximate) solutions to the governing equations with only end loading. Loading applied along the length of the element is taken into account by using equivalent end loads.

Four beam element types are shown in Fig. 5.15.

- The *plane frame element* (also called the *2D beam element*) incorporates a single plane of bending plus axial effects with three degrees of freedom per node – Fig. 5.15(a).
- The *grillage element* incorporates a single plane of bending and torsional effects – Fig. 5.15(b).
- The *bar element* incorporates only axial action – Fig. 5.15(c).
- *Beam element in three dimensions* (also called the *3D beam element*) is the most general type of line element, incorporating bending in two planes, axial and torsional actions with six degrees of freedom at each node – Fig. 5.15(d).

In all cases, the bending component may include shear deformation (thick beam) or neglect shear deformation (thin beam).

5.5.3   Higher-order beam elements

Higher-order beam elements (which do need mesh refinement), with more than two nodes, are used to model situations such as curved beams (Section 5.8.2), cross-sectional variation (Section 5.8.3) and non-linear geometry effects (Section 10.3.5).

5.6   Connections

In this section, modelling issues for connections in skeletal frames are discussed.

5.6.1   Basic connection types
Pin connections

The basic assumption for a pin connection is that the parts of the structure being connected are free to rotate independently. Therefore, no moment is transferred between the connected parts. A common type of pin connection is a clevis arrangement, as shown in Fig. 5.16.

Moment connections

The basic assumption for a fully rigid moment connection is that the relative angle between the axes of the connected parts does not change. For example, Fig. 5.17 shows two members connected at an angle θ. To have a fully rigid moment connection, the angle θ does not alter under loading. The connection can rotate, but θ will not change. This is denoted as 'full continuity' between the members.

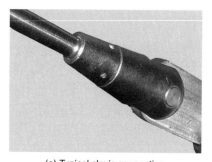

(a) Typical clevis connection

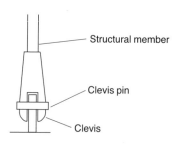

(b) Section showing clevis pin

Figure 5.16   Clevis connection.

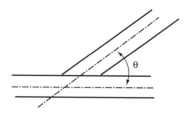

Figure 5.17   Fully rigid connection.

Figure 5.18 Beam-to-column moment connection.

As an example, Fig. 5.18 shows a typical beam-to-column connection in steelwork, which would be treated as being a fully rigid moment connection. To transmit a moment there must be a lever arm within the connection that can provide an internal moment. A main feature of the connection in Fig. 5.18 is that the flanges of the beam are connected to the flange of the column to make it capable of transferring a full moment. Note the bolts on the beam end plate above and below the beam – this is a clear indication that it is designed as a moment connection.

Semi-rigid moment connections

In many cases the true behaviour will lie between that of a pin and a continuous connection. For example, at a beam-to-column connection in steelwork there may be a relative rotation, ϕ, at the end of a beam relative to the column – shown exaggerated in Fig. 5.19. Such relative rotation will cause a reduction in the support moment and, consequently, an increase in mid-span moment as compared with a rigid joint condition. The semi-rigid connection can be modelled using a joint element at the connection – see Fig. 5.23(d).

A fundamental principle is that if the analysis predicts a moment at the connection then the connection must be at least strong enough to take that moment. An analysis is carried out using assumptions about the rotational rigidity of the connections, and the connection is designed for the resulting internal force actions.

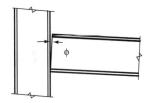

Figure 5.19 Local rotation between a beam and a column.

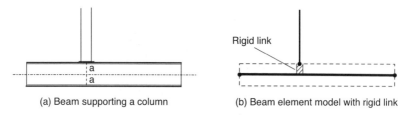

(a) Beam supporting a column   (b) Beam element model with rigid link

Figure 5.20 The use of a rigid link used to model the finite width of a beam.

The lower bound theorem is used to justify this procedure. Note, however, the reservations about use of the lower bound theorem given in Section 2.3.3.

5.6.2   Treatment of the finite depth of a beam using rigid links

In this section, the rationale for modelling the finite depth of a beam as a rigid link is discussed.

Beam elements are defined along a line, but sometimes it is necessary to take account of the finite depth of the beam to model more accurately the effect of the connection or a support. Figure 5.20(a) shows a beam with a fairly high span-to-depth ratio which supports a column. The line *a–a* is straight and at right angles to the axis of the beam at the column position. The fundamental assumption for the ETB is that such a line will remain straight after deformation – see Fig. 5.2.

Figure 5.20(b) shows a beam element model of the system. The beam elements are defined at the centroidal axes of the members. To take account of the fact that the column is connected at the edge rather than at the centre of the beam, a rigid link is inserted as shown. This forces the deformation between the centroidal axis of the beam and the connection to the column to remain a straight line, as is assumed for the ETB.

The rigid link can be implemented in a number of ways, as discussed in Section 4.5.2. The simplest method is to use an element which includes a rigid link. The use of rigid links is discussed in Sections 5.6.3 and 6.3.4.

5.6.3   Modelling beam-to-column connections in steelwork
Pin connections

In Fig. 5.21 the diagonal member has two bolts, and although this connection can take a moment it is very unlikely that it was designed to do so. The connection of the post with four bolts has better potential to take a design moment, but again it is unlikely that it would have been so designed.

Figure 5.22 shows a beam-to-column flange connection with only a web cleat. This can also take some moment, but because the flanges of the beam are not connected to the column, the connection could be neither stiff enough nor strong enough to be treated as a moment connection. It would have to be treated as a pin connection.

The ends of bar elements are, by definition, pinned. Other standard element connections assume full continuity at the nodes. To create a pin at the end of a

Figure 5.21   Connection in a steel truss.

beam element, an end release is needed. Beam elements normally allow this as an option.

Moment connections

Figure 5.23(a) shows a moment connection between a steel beam and a column. Figure 5.23(b) shows the conventional model of such a connection, where the beam and the column are considered to be flexible to their intersection point. Neglecting the effect of the finite sizes of the members will tend to underestimate the stiffness of the connection. It would be conservative to take the design support moment for the beam at the intersection point, but it is normal to take it at the face of the column.

Figure 5.22   Steelwork connection with only a web cleat.

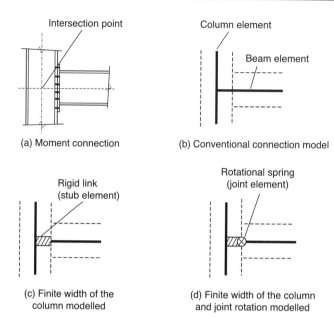

Figure 5.23 Models for beam-to-column moment connections.

Figure 5.23(c) shows how the effect of the finite width of the column can be modelled using a rigid link – known as a *stub element* – between the intersection point and the face of the column (Section 5.6.2).

Figure 5.23(d) shows the same model as Fig. 5.23(c) but with a rotational spring between the end of the stub and the beam to take account of the rotational stiffness of the connection. This can be implemented using a joint element (Section 4.7.5)).

Appendix J of Eurocode 3 (CEN 2004b) gives rules for estimating values of the rotational stiffness of steelwork connections.

The inclusion of the rigid links will increase the effective stiffness of the beam, whereas the rotational spring will decrease the stiffness. The conventional model in Fig. 5.23(b) may therefore give reasonable predictions of stiffness due to compensating assumptions. Whether or not the rigid links and the rotational springs are important can be assessed using a sensitivity analysis (Section 2.4.4).

Pin connections

Figure 5.24(a) shows a beam-to-column connection with a web cleat, which would be modelled as a pin. Figure 5.24(b) shows the pin at the centre-line of the column, and 5.24(c) shows the pin at the end of the stub element. The model in Fig. 5.24(c) is clearly the more accurate and will automatically take account of the effect of the eccentricity of the beam support on the column.

Modelling the connections in pin-jointed frames

To model a pin-jointed frame, either bar elements (which do not have rotational freedoms) or beam elements with end releases are used. If the latter approach is adopted then the following rules must be followed.

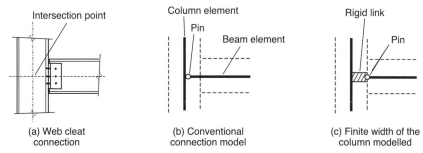

| (a) Web cleat connection | (b) Conventional connection model | (c) Finite width of the column modelled |

Figure 5.24 Models for pinned beam-to-column connections.

1. If *m* members meet at a joint then only *m* − 1 members' ends should be released. Figure 5.25 shows an example of this for a beam-to-column connection. Three members meet at the joint but only two releases should be applied. This is because with beam elements the analysis program defines rotational freedoms at the nodes, and if no rotational stiffness is defined at a node then the equation solver will fail. The single unreleased member end provides a rotational stiffness, but since there is no other element with bending stiffness to connect to it, the joint behaves as a pin.

2. With space frames, the torsional freedom (i.e. the rotational freedom about the longitudinal axis – the local *x* axis) at only one end of a member can be released. If the torsional freedoms at both ends are released then the element will be free to rotate about its longitudinal axis and the solver will fail.

5.6.4 Connections in concrete

Figure 5.26 shows a reinforced concrete beam-to-column connection that is detailed to take hogging moment. This would be treated as a moment connection in the model, provided there can be no sagging moment in the beam at the connection. The top reinforcement in the beam must be properly anchored in order to transfer a moment. If there can be sagging moment then the bottom main steel must also be anchored to the column.

Precast concrete systems tend to have special connections which behave as pins. If a pin connection is assumed for a cast-in-situ system, it should be ensured that cracking at such locations will not be a problem.

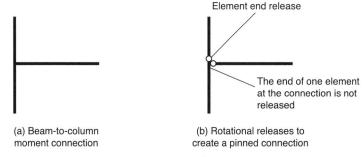

| (a) Beam-to-column moment connection | (b) Rotational releases to create a pinned connection |

Figure 5.25 Creating a pin connection at a beam-to-column joint.

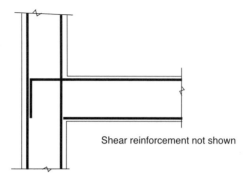

Shear reinforcement not shown

Figure 5.26 Moment connection for a reinforced concrete beam-to-column connection.

5.6.5   Eccentricity of members at a joint

Figure 5.27(a) shows the intersection of a beam, a column and two diagonal members (details of connections are not shown). The axis of member A passes through the intersection of the axes of the column and the beam, and there is therefore no eccentricity for these three members. However, the axis of member B has an eccentricity, e, from the intersection point. The axial forces in the members will tend not to be significantly affected by connection eccentricity, but if the axial force in the diagonal member is n_d, then there will be a moment

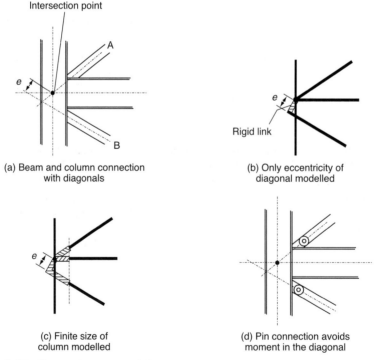

Figure 5.27 Connection eccentricity.

$= n_d e$ applied to the joint. It is important to ensure that this moment does not adversely affect the performance of the system. Ignoring this effect has contributed to at least one major collapse – that of the Hartford Civic Center roof (Levy and Salvadori 1992), see Section 3.8.2.

It is normally recommended that eccentricities of the type shown in Fig. 5.27 should be avoided, but this can result in awkward, and in some cases unsatisfactory, detailing. If in doubt about the effect of the eccentricity, it should be included in the analysis model or a sensitivity analysis should be carried out.

Modelling eccentricities

The eccentricity can be modelled using a rigid offset as follows.

1. Decide on a position for the node that will represent the intersection point – for example at the point where most of the axes do intersect (typically at the beam–column or chord–post intersections).
2. Provide rigid links for those members that are eccentric to the node, as illustrated in Fig. 5.27(b). The modelling of rigid links is described in Section 4.5.2.

Figure 5.27(c) shows how the rigid offsets can be used to model the eccentricity of the diagonals and the finite size of the column (see also Section 5.6.2).

It may be important to ensure that little or none of the eccentric moment is applied to the diagonal element since it is not normally designed to resist bending moments. This may be achieved by:

- detailing the connections at the ends of the diagonal member as pins (Fig. 5.27(d)) (this will prevent moment being transferred to the diagonal members)
- ensuring that the rotational stiffnesses of the members to which the diagonal member is connected are significantly greater than that of the diagonal member itself. This will mean that the moment due to the eccentricity will be taken mainly by the other members.

Modelling scaffolding systems

The members of scaffolding systems tend not to be connected in a single plane. Figure 5.28 shows a diagrammatic connection for scaffolding. There is eccentricity both in the plane and out of the plane. A particular problem with this type of system is that there is low rotational stiffness at the connections and therefore the proportion of the eccentric moment taken by the diagonal, for example, may not be negligible. The effects of this are to reduce the effective stiffness of and to increase the stress in the diagonals. To model the behaviour accurately, the

Figure 5.28 Connection for scaffolding.

eccentricities and rotational stiffnesses at the joint need to be modelled. Getting correlation between measurements and the results of an analysis model of such systems can be very difficult.

5.7   Distribution of load in skeletal frames

5.7.1   Vertical load in beam systems

Understanding how load is distributed in a structure is an essential feature in modelling. In this section, a classification of how vertical load may be distributed is given.

Flexible beam on rigid supports

It is common to assume that a uniformly distributed load on a beam (or on a one-way spanning slab) will distribute equally to either support. This is correct for simply supported spans but is approximate for continuous spans.

Figure 5.29(a) shows two beams of equal spans, both simply supported on walls and each taking a uniformly distributed vertical load of W. Each support takes half of the load from each span to give the central reaction as double the external reaction. Figure 5.29(b) shows the same system but with the beam continuous over the central support. The central support now takes $5/4W$ and the ends supports $3/8W$ each. The continuity at the support tends to draw the load towards the support. The distributions of Fig. 5.29 represent the flexible beam on rigid support model of beam behaviour, which is normally valid for slabs spanning onto walls or for beams spanning onto columns or walls.

The distribution of Fig. 5.29(a) is often used as a first approximation, whether or not there is continuity at the supports. In a particular context the validity of doing this may need to be assessed.

Flexible beam on flexible supports

When a continuous beam is supported on other beams, as in Fig. 5.30(a), the load distribution will be affected by the stiffness of the support system. There is no simplified approach to distributing load in such cases. A grillage model (Section 5.12) or a 3D model is needed.

Rigid beam on flexible supports

Figure 5.30(b) shows this situation. This situation is relevant, for example, to the distribution of lateral load in buildings and to the axial forces in the column supports for walls. In this case, if the beam does not rotate, the load is distributed

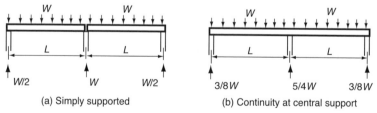

Figure 5.29   Distribution of load by a flexible beam system with rigid supports.

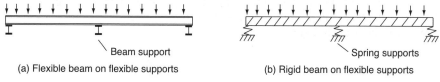

Beam support

(a) Flexible beam on flexible supports

Spring supports

(b) Rigid beam on flexible supports

Figure 5.30 Beams on flexible supports.

to the supports in proportion to their stiffnesses. Beam rotation can also be considered – the rigid plate on springs model (MacLeod 1990).

5.7.2   Distribution of lateral load
An example of the distribution of lateral load in a building structure is given in Section 12.2.

5.8   Modelling curved and non-uniform members
5.8.1   Curved members
Curved members can be modelled using:

- a mesh of straight elements
- elements which are themselves curved.

The following section demonstrates that the latter strategy is the better option.

5.8.2   Case study – modelling of curved beams
Figure 5.31 shows an analysis model of a curved beam. It is a quarter of an arc of a circle of radius 3.0 m. The lower end is fully fixed and the upper end has two separate loading conditions, namely:

- *in-plane case* – 10.0 kN applied vertically in the plane of the beam, as shown in Fig. 5.31
- *out-of-plane case* – 0.1 kN applied at right angles to the plane of the beam, i.e. in the z direction.

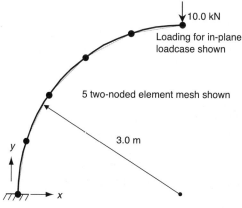

10.0 kN

Loading for in-plane
loadcase shown

5 two-noded element mesh shown

3.0 m

y

x

Figure 5.31 Model of a curved beam.

Table 5.1 Benchmark values

Loadcase	Δ_{bm}: m
In-plane	0.01834840864
Out-of-plane	0.07997322052

Element types

Two types of beam element were used.

- *Two-noded* (LUSAS BMS3[1]) – straight thick 3D beam element, with shear deformation neglected in this case.
- *Three-noded* (LUSAS BS3[1]) – thin beam 3D element, which can model curved lines and has mid-length nodes.

The section of the beam is a steel $254 \times 146 \times 37$ mm universal beam, bending about its major axis for the in-plane case ($E = 209$ kN/m^2) and about its minor axis for the out-of-plane case.

Indicative parameter

The deflection in the line of the load, Δ_{tip}, is taken as the indicative parameter. In Fig. 5.32 the ordinate of the diagrams is the percentage difference from the benchmark value, i.e.

$$\Delta^* = (\Delta_{tip} - \Delta_{bm})/\Delta_{bm} \tag{5.15}$$

where Δ_{bm} is the benchmark value.

Benchmark values

The benchmark values for deflection, given in Table 5.1, were from the results of a 100-element mesh of three-noded elements.

Results

Figures 5.32(a) and (b) are plots of Δ^* against the number of elements in the mesh. The convergence curve for the two-noded elements starts to flatten off at around the eight-element mesh, whereas for the three-noded element even a two-element mesh gives a good result. It is clear that if three-noded elements that take account of the curved shape are available, they should be used for curved shapes.

Validation analysis

The graphs of Figs 5.32(a) and (b) show the potential accuracy of the computational model, i.e. the use of several elements which are either straight or curved to model a circular curve. The conceptual model (i.e. ETB) for the in-plane loading is likely to be adequate for practical purposes in this case. However, comparison with measured results for the out-of-plane loading could show the benchmark values quoted in Table 5.1 for the out-of-plane case to be significantly less than

[1] LUSAS finite element modeller, FEA Ltd.

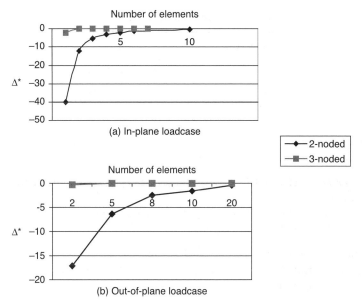

Figure 5.32 Convergence analysis of a curved beam.

a measured value. This is because with out-of-plane action there will be significant bending torsion in the open I section. This effect is not included in the elements used for the results of Fig. 5.32(b). For better accuracy in the out-of-plane case an element which includes bending torsion should be used (Section 5.4.3).

5.8.3 Modelling members with non-uniform cross section
Members with non-uniform cross section can be modelled using:

- a mesh of uniform elements where the properties of the element are taken to be those at the centre of the element
- elements that take account of variation of cross-sectional geometry.

Case study 5.8.4 shows that the latter strategy is the better option.

5.8.4 Case study – tapered cantilever
Context
Figure 5.33(a) shows a tapered cantilever beam. The thickness of the beam is constant 100 mm and the material is concrete ($E = 24.0\,\text{kN/mm}^2$). The uniform taper is from 200 mm depth at the fixed end to 100 mm at the tip. One end is fully fixed and at the other end a vertical load of 10.0 kN is applied.

Element types
The 2D equivalents of the element types used for the curved beam study of Section 5.8.2 were used. The two-noded elements have constant cross section, and to model the variation in cross section the member is divided into elements of equal length, having values of area (A) and second moment of area (I) corresponding to the

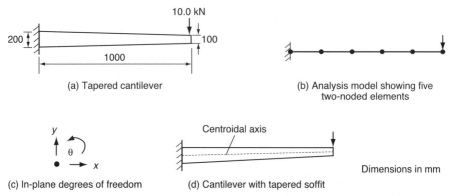

Figure 5.33 *Tapered cantilever.*

section at the mid point of each element. For the three-noded elements the cross section parameters at each element node are part of the data.

Figure 5.33(b) shows a typical 2D in-plane analysis model of the cantilever with five two-noded elements.

Indicative parameter

The deflection in the line of the applied load is used as the indicative parameter – characterised by Δ^* as defined by equation (5.15).

Benchmark value

The benchmark value used is that from the run with a ten-element mesh of three-noded elements – $\Delta_{bm} = 3.407337$ mm.

Results

Figure 5.34 shows the results of the convergence analysis.

- The rate of convergence for the three-noded element is significantly superior to that for the two-noded element. A single three-noded element gives the same order of accuracy as a mesh of five two-noded elements.
- The two-noded results converge monotonically from above – i.e. the results are all greater than the benchmark. For the three-noded element the results converge monotonically from below.
- The member used has a shallow taper – conclusions from this study may not be valid for steeper tapers.

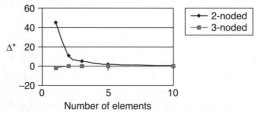

Figure 5.34 *Convergence analysis for the tapered cantilever.*

(a) Haunched beam   (b) Analysis model

Figure 5.35 Haunched concrete beam.

5.8.5   Cantilever with a tapered soffit

Figure 5.33(d) shows the more common form of a cantilever, one with a tapered soffit. In this case there will be an axial load in the elements, but the effect on the tip deflection is negligible for this taper (tip deflection is 0.12% greater than for the uniform taper situation in this context).

5.8.6   Haunched beams

Figure 5.35(a) shows a haunched beam as part of a concrete frame. This can be modelled with tapered sections at each end. The slope of the elements used to model the haunch (Fig. 5.35(b)) will cause a sort of end arch action to develop, which may be structurally advantageous if there is lateral restraint to outward movement at the ends of the beam.

5.9   Triangulated frames

5.9.1   Modelling issues

Triangulated frames tend to resist loading mainly by axial forces in the members. When there are no moment connections, the frame is treated as being *pin-jointed*.

However, if the system does have moment continuity at the connections (Fig. 5.36 shows a triangulated bridge truss with moment connections) it is essential that the resulting moments in the members are taken into account in the member sizing process (see Section 2.2.6). A sensitivity analysis (Section 2.4.4) should be carried out if there is doubt about the effect of the moment connections.

Figure 5.36 Triangulated bridge structure with moment connections.

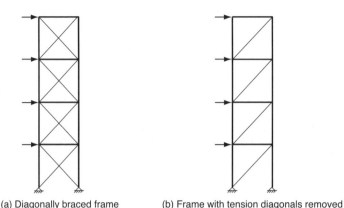

(a) Diagonally braced frame   (b) Frame with tension diagonals removed

Figure 5.37 Removal of compression diagonals.

5.9.2 Euler buckling effect of members

The need to take account of the second-order Euler buckling effect of members of a triangulated frame can be tested using the $N/N_{cr} < 0.1$ criterion set out in Section 10.3.3. If this effect needs to be taken into account then a non-linear geometry analysis can be carried out (Section 10.2). Where a bracing member is not expected to take compressive load (e.g. if a flat bar member has been specified) it should be made ineffective when the load is compressive. Such a member is denoted here as a *nil compression member*. Some software systems can cater for nil compression members automatically, but if such a facility is not available, the following algorithm can be used.

1. Run the frame with all members.
2. Identify the nil compression elements that have compressive axial force and remove them from the model (or make them ineffective if this facility is available in the software).
3. Re-run the model to make sure that the redistribution of load caused by removing elements has not caused further nil compression elements to go into compression.
4. Repeat until no nil compression elements are in compression.

Note that there is a possibility in the above cycle that a removed nil compression element could go back to tension.

Figure 5.37(a) shows a model of a diagonally braced frame in a building. If the diagonals are not designed to take compression then the model of Fig. 5.37(b) needs to be used. In this situation the compression diagonals can be identified by inspection. See the example in Section 12.2.

5.10   Parallel chord trusses

5.10.1   General

A parallel chord truss (Fig. 5.38) can be modelled as an equivalent beam:

- to treat the truss as a single element in order to reduce the number of elements to be considered directly in an analysis model – see Fig. 5.39

Figure 5.38   Parallel chord truss bridge structure.

- as a checking model
- as an aid to understanding of behaviour.

5.10.2   Definitions

The term *chord* refers to the members of a truss that are in the direction of the span – normally the horizontal members. *Post* describes the members of a truss that are at right angles to the chord. *Web members* are the diagonal members and the posts.

A diagonally braced frame in a building represent a type of parallel chord truss – see Fig. 5.37.

5.10.3   Behaviour

Parallel chord trusses tend to behave in a similar way to an I-section beam, and in order to understand their behaviour it is necessary to distinguish between bending mode deformation and shear mode deformation of a beam.

Bending mode deformation is characterised by equation (5.1). Figure 5.40(a) shows a cantilever with end point load and the corresponding cubic curve for displacement due to bending mode deformation. This results from equation (5.1), which needs to be integrated twice to get the displacement, v. If M is linear then v will be cubic.

Shear mode deformation is characterised by equation (5.3). Figure 5.40(b) shows a cantilever with two loadcases – a point load and a uniformly distributed load. For the point loadcase, equation (5.3) is integrated once to get v, and since S is constant, v is linear. With the uniformly distributed loadcase, S is linear and

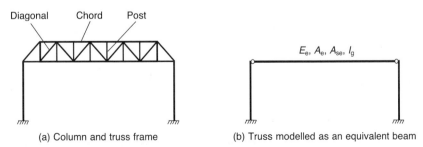

(a) Column and truss frame   (b) Truss modelled as an equivalent beam

Figure 5.39   Truss treated as an equivalent beam.

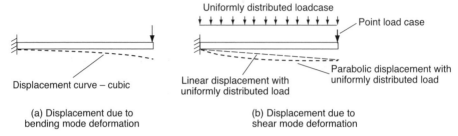

Figure 5.40 Displaced shapes of a cantilever with bending mode and shear mode deformation.

therefore v is parabolic. Note, however, that the parabolic shape is different from bending mode in that the maximum slope of the curve is at the fixed support (because the shear is maximum there) whereas the slope of the curve there is zero under bending mode.

For the parallel chord truss there is a bending mode component of deformation due to the axial deformation of the chords and a shear mode component of deformation due to the axial deformation of the diagonals and the posts. The chords act like the flanges of an I-section beam, and diagonals and posts simulate web action. The correspondence between the components are set out in Table 5.2. A main difference is that for I beams, bending is normally the dominant component of deformation whereas shear mode deformation tends to be dominant for parallel chord trusses.

5.10.4 Equivalent beam model
Basic properties of the equivalent beam
The properties to be established are:

- I_g – the equivalent I value
- K_{st} – the shear stiffness.

Equivalent I value, I_g
The basic process is to treat the chord areas as a cross section of a member in bending. Calculate the position of the centroid of the areas of the chords and calculate I_g (Fig. 5.41(a)). The I values of the areas about their own centroids can be ignored in the calculation.

Table 5.2 Comparison of contributions to deformation components

System	Dominant source of deformation component	
	Bending	Shear
I beam	Flanges	Web
Parallel chord truss	Chords	Diagonal and post members

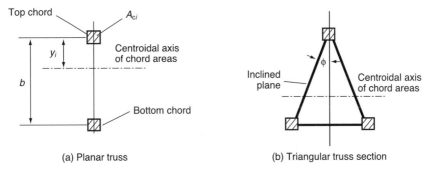

(a) Planar truss   (b) Triangular truss section

Figure 5.41   Parameters for the equivalent I value.

Thus

$$I_g = \sum A_{ci} y_i^2 \tag{5.16}$$

where A_{ci} is the area of chord member i and y_i is the distance from the centroidal axis of the chord areas to the centroid of the area of chord i.

For a plane truss with the same members for the top and bottom chords (Fig. 5.41(a)), the equivalent I value is

$$I_g = A_c b^2 \tag{5.17}$$

where A_c is the area of a chord and b is the depth of the truss.

Shear stiffness, K_{st}

For a plane triangulated truss the shear stiffness K_{st} can be estimated as the reciprocal sum of the contributions from the axial deformation of the diagonals and the axial deformation of the posts, i.e.

$$K_{st} = \left[\frac{1}{1/(f\, E_d A_d \sin^2 \theta \cos \theta) + 1/(A_p E_p \cot \theta)} \right] \tag{5.18}$$

where f is a factor ($= 1.0$ for single bracing, $= 2.0$ for cross bracing, provided that the compression diagonals are effective, and $= 0.5$ for K bracing), E_d is the Young's modulus for the diagonal member, A_d is the area of a diagonal member, θ is the angle between the chord and the diagonal, A_p is the area of a post and E_p is Young's modulus for the post.

The flexibility of the posts is often neglected for singly braced trusses and should be omitted for cross-braced trusses. Neglecting the flexibility of the posts, equation (5.18) reduces to

$$K_{st} = f E_d A_d \sin^2 \theta \cos \theta \tag{5.19}$$

If the truss has a triangular cross section (Fig. 5.41(b)), then the shear stiffness for one of the inclined planes $K_{s,planar}$ should be calculated. The shear stiffness for the truss is then

$$K_{st} = 2 K_{s,planar} \cos^2 \phi \tag{5.20}$$

where ϕ is the angle between the plane of bending and the inclined planes (Fig. 5.41(b)).

Validation information for the equivalent beam for parallel chord trusses

- The equivalent beam as described in this section can give results that are quite close to those from a full 2D or 3D model. It can be useful for checking and for preliminary design but it may not have adequate accuracy for final design.
- The equivalent beam assumes that the truss is uniform along its length.
- Equation (5.16)
  - The use of equation (5.16) will tend to provide fairly accurate estimates of bending mode deflection and stiffness for frames with singly braced panels.
  - With cross-braced panels, the bracing will also contribute to the bending mode stiffness, and therefore ignoring it in the definition of I_g will tend to overestimate the deflection. This will normally be conservative.
- Equation (5.18)
  - The use of equation (5.18) for the shear mode stiffness will also tend to give fairly accurate estimates of bending mode deflection and stiffness.
  - Equation (5.18) neglects the effect of moment connections, leading to an overestimate of deflection if there are moment connections. An estimate of the effect on deflection of moment connections between the chords and the posts can be made by using a combined shear stiffness of

$$K_{sc} = K_{st} + K_{sv} \tag{5.21}$$

where K_{sv} is obtained from equation (5.24).

Using the equivalent beam for checking transverse deflection of a plane truss
The equivalent frame can be used as a checking model for transverse deflection by adding the bending mode and shear mode components together, i.e.

$$\Delta_{frame} = \Delta_b + \Delta_s \tag{5.22}$$

where Δ_{frame} is the deflection of the frame, Δ_b is the component due to bending mode deformation (i.e. due to axial deformation of the chords in this situation) and Δ_s is the component due to shear mode deformation (i.e. due to axial deformation of the diagonals and posts).

Both Δ_b and Δ_s can be calculated using the formulae given in Table A4, using equation (5.16) for the I value and equation (5.18) for K_{st}.

Properties of the equivalent beam for use in a frame model
This section indicates how to calculate the member properties for an equivalent beam in a frame element model (Fig. 5.39).

Shear deformation needs to be included and therefore a 'thick beam' element must be used.

Plane frame model (2D)
The properties of the equivalent beam may be established as follows.

- E *value* – use E_c, the E value for the chord members.
- Poisson's ratio, ν – any arbitrary value can be used since it is not a parameter of K_{st}, the shear stiffness of the equivalent beam. Use, for example, $\nu = 0.25$.

- *Area* – use $A_e = \sum A_{ci}$, the sum of the cross-sectional areas of the chord members.
- I *value* – use I_g from equation (5.16).
- *Shear area, A_{se}* – use

$$A_{se} = \frac{2(1+\nu)}{E_c} K_{st} \tag{5.23}$$

where K_{st} is calculated using equations (5.18) or (5.19).

3D frame model

The properties of the equivalent beam may be established as follows.

- Calculate E, ν and A_e as for the planar truss.
- Calculate I_g and A_{se} as for the planar truss but for both planes of bending – use equation (5.20) as appropriate.
- *Torsional constant* – if the torsional stiffness of the truss is not important any small J value, such as the J value of a chord member, can be used. If the torsional stiffness of the system is considered to be important then a full 3D element model should be used.

Using the equivalent beam for assessing the effect of member sizes on frame stiffness

Equation (5.22) can be used to carry out sensitivity analysis in relation to frame stiffness. This can be done by varying the numerical values of parameters or by differentiating equation (5.22) to find what will tend to give the greatest rate of change.

An example of this process is given in Section 12.1.7.

Using the equivalent beam for estimating member forces in parallel chord trusses

The maximum axial forces in the members of a parallel chord truss can be estimated as follows.

1. Find the maximum moment M_{max} and the maximum shear force S_{max} in the equivalent beam.
2. Calculate the maximum chord force N_c by using $N_c = M_{max}/b$, where b is the depth of the truss (centre-to-centre distance between the chords).
3. Calculate the maximum diagonal force n_d by using $n_d = S_{max}/\sin\theta$ for single diagonals or $N_d = S_{max}/(2\sin\theta)$ for cross bracing (provided that the compression diagonals have adequate bending stiffness), where θ is the angle between the chord and the diagonal.

5.11   Vierendeel frames

5.11.1   Definitions

The term *vierendeel frame* refers to a frame made up of rectangular panels with no diagonal bracing – see Fig. 12.1. Such frames derive lateral stiffness mainly from

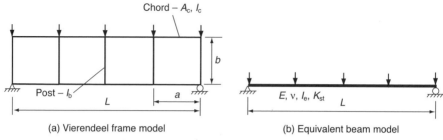

(a) Vierendeel frame model   (b) Equivalent beam model

Figure 5.42   Equivalent beam model of a vierendeel frame.

bending of the members made possible by moment connections between them. Such frames are also called *rigid jointed frames*, but *vierendeel* is a better name because triangulated frames can also have rigid joints – see Fig. 5.36.

Vierendeel frames tend to be much less stiff than triangulated frames but provide clearer spaces.

The concepts of a *chord* and a *post* for a vierendeel bridge-type frame are shown in Fig. 5.42. For a vierendeel frame resisting lateral load in a building, a 'column' is equivalent to a 'chord' and a 'beam' is equivalent to a 'post'.

5.11.2   Behaviour

The analogy between a vierendeel frame and a beam is similar to that for a parallel chord truss except that the shear mode component is due to bending of the chords and the posts rather than to axial deformation of the diagonals and posts. The case study in Section 12.1 illustrates the behaviour of a vierendeel frame.

5.11.3   Equivalent beam model
Equivalent I value, I_g
As for parallel chord truss, equation (5.16).

Shear stiffness, K_{sv}

$$K_{sv} = \frac{24E_cI_c}{a^2[1 + 2\psi]} \tag{5.24}$$

where I_c is the *I* value of a chord member, a is the length of a panel of a chord (Fig. 5.42(a)), $\psi = (I_c/a)/(I_p/b)$, I_p is the *I* value for a post and b is the depth of the frame.

Validation information for the equivalent beam for vierendeel frames

- *Equivalent I value* – as for parallel chord trusses, Section 5.10.4.
- The equivalent beam assumes that the frame is uniform along its length.
- *Equation (5.24)* – the main assumption for equation (5.24) is that there are points of contraflexure in the mid-lengths of all the chords and posts. This depends on the following.

- ○ The main parameter for the chord points of contraflexure is the ratio of chord bending stiffness to that of the beams, i.e. $\psi = (I_c/a)/(I_b/b)$. As ψ decreases below 1.0 (beams stiffer than the columns) the 'shear beam' situation is approached, where the posts can be considered to be rigid in bending in relation to the chords. In such a situation (typically $\psi < 0.1$), the points of contraflexure will be close to mid-length of the chord members.
  - ○ As ψ increases above 1.0, the accuracy declines, with the points of contraflexure moving away from the mid-length positions, particularly in the centre panels and the end panels of the frame.
  - ○ The points of contraflexure will be at mid-length of the posts only if the top and bottom chords have the same cross section.
- • Imposing pins at mid-length of the members is equivalent to the application of releases to the system and therefore equation (5.24) will tend to underestimate the stiffness of the system, i.e. deflections due to shear mode deformation of the equivalent beam will tend to be overestimated.
- • The accuracy tends to improve as the number of panels in the frame increases, i.e. accuracy tends to be low for frames with a small number of panels – see Section 12.1.6.

5.12   Grillage models

A grillage is normally (but not always) a system of horizontal beams which spreads vertical loading to supports (Fig. 5.43). Grillages are commonly used for bridge decks and machinery support structures. To be effective in distributing the load between the main beams, the cross beams must be continuous or have moment connections to the main beams. Torsion may be important in the load transfer and may therefore need to be considered in the design of the members.

A grillage can be modelled using a grillage element (Fig. 5.15(b)) or by using 3D beam elements.

The grillage element has bending about the local y axis, which combines with the shear in the z direction to give bending in the xz plane. In addition, there is torsion about the local x axis (Fig. 5.15(b)).

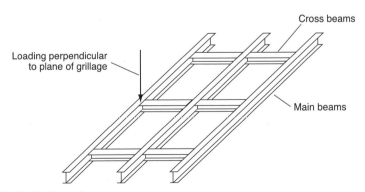

Figure 5.43 Grillage frame.

The effect of torsion may be negligible, but it is important that the validation information for torsion discussed in Section 5.4.5 should be considered.

5.13   3D models

To validate the use of 3D beam elements the validation information in Sections 5.2.7 (bending), 5.3.3 (axial effects) and 5.4.5 (torsion) need to be considered. An example of the use of a 3D model is given in Section 12.2.

5.14   Plastic collapse of frames

5.14.1   Prediction of collapse loads – limit analysis

The traditional approach to estimating the plastic collapse load of a rigid jointed frame (based on the original work by Baker *et al.* 1951) is to postulate a set of failure mechanisms (e.g. Fig. 5.44(b)) and calculate the corresponding collapse loads. The lowest collapse load is then used as the collapse load for the system. This has the following disadvantages.

- It can be difficult to ensure that the real collapse mechanism has been identified, and because any collapse mechanism gives an upper bound to the collapse load, this is not a safe bound.
- The loading has to be proportional, i.e. the vertical and lateral loads must increase in proportion to each other. This is not the case in real situations; vertical load is usually applied first, then lateral load. In non-linear analysis, the order of application of loads is important, i.e. *superposition does not apply* (Section 2.3). Therefore, the use of proportional loading introduces inaccuracy to the solution.
- As the collapse load is approached, non-linear geometry effects may become important.

These limitations can be avoided by using modern software.

5.14.2   Prediction of plastic collapse using an iterated elastic analysis

The following process can be used to predict plastic collapse of frameworks using an elastic analysis program.

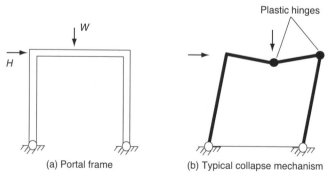

(a) Portal frame   (b) Typical collapse mechanism

Figure 5.44   Plastic collapse of a portal frame.

1. A first increment of load is applied and the bending moments in the frame are searched to find the highest value.
2. The applied load is factored to make a plastic hinge at this location.
3. A pin is inserted at the plastic hinge location.
4. A new increment of load is applied with a plastic moment (pair of equal and opposite restoring moments) acting at the new pin. The results from this increment are added to the internal forces from the factored first increment results and the location of the next plastic hinge is identified.
5. The second increment is factored to just reach the second plastic hinge.
6. The process is repeated in load increments until a mechanism condition is reached (e.g. when the solution process fails to converge). The true collapse mechanism has now been found.

This is the basic process. Using specialist software (such as Fastrak)[2], features such as closing joints can be taken into account (Davies 2001).

5.14.3   Prediction of plastic collapse using a finite element solution
The collapse load can be found using a finite element package by the following process.

1. Set up the frame model with joint elements at all locations where plastic hinges may occur. The joint elements must be capable of modelling plastic behaviour. Appropriate plastic moments are defined for each joint element.
2. Set up the required loading regime.
3. Run as a non-linear analysis, including non-linear geometry if appropriate.

This is considered to be advanced analysis, and detailed consideration is outside the scope of this text.

5.14.4   Validation information
Plastic collapse modelling of frames

- *Sections are assumed to be capable of developing full plastic moments* – issues in this respect include the following.
  - *Ductility* – for the development of a plastic hinge in a steel section the material of the beam will need sufficient ductility to allow the whole section to become plastic. The ductility of steel is normally adequate for this.
  - *Local buckling* – while a rectangular beam can develop a full plastic moment, other section shapes can experience local buckling effects before the full plastic condition has been reached. British Standard BS 5950-1 for the design of steel structures (BSI 2000) classifies steel sections with respect to their ability to generate full plastic moments and provides rules for estimated plastic capacity.
  - *Axial and shear forces* – can affect the ability of a section to mobilise the full plastic moment, although these effects are not often important.

---

[2] Fastrak Portal Frame Design Software, CSC Ltd.

- *Strain hardening* (Section 7.3.1) – this is neglected. It will be conservative, but can be included in the model.
- *Non-linear geometry effects* – these effects may be important.
  - Lateral torsional buckling may affect the beams. This is normally taken into account in the design rules for the beams.
  - Second-order effects in the columns may not be negligible – see Chapter 10.

6 Plates in bending and slabs

6.1 Introduction
This chapter discusses issues in relation to planar structural systems that support load at right angles to the plane. This applies particularly to floor slabs and bridge decks. The application to steel plates is not dealt with specifically, but, for example, the modelling techniques described for beam supported slabs are applicable to the modelling of stiffened steel plates.

6.2 Plate bending elements
Plate bending element are used:

- to model flat plates that are subject only to out-of-plane bending actions
- as a basic component of a traditional flat shell element.

This section gives general information about modelling and basic theory for thin plate bending.

6.2.1 Plate bending element basics
Degrees of freedom
Figure 6.1(a) shows a typical quadrilateral plate bending element. The conventional freedoms at a node in a plate bending element mesh are one translational freedom normal to the plane of the plate plus two rotational freedoms acting about the axes in the plane of the plate (Fig. 6.1(b)).

Constitutive relationships
The normal plate bending element uses thin plate bending theory, sometimes described as the Kirchhoff plate bending theory. The basic assumption of this

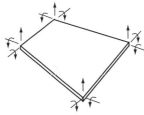

(a) A quadrilateral plate bending element	(b) Degrees of freedom for plate bending elements

Figure 6.1 Plate bending element.

theory is that normals to the plane of the plate in the undeformed state remain straight in the deformed state. This is the two-dimensional equivalent of the 'plane sections remain plane' assumption for uniaxial bending (Section 5.2.2).

Some elements use constitutive relationships, which include shear deformation (*thick plate* elements).

Element types
Plate bending elements tend to be triangular or rectangular, with or without mid-side nodes, as for plane stress elements (Section 4.2.6). The assumptions in their derivation tend to be more complex than for plane stress, and their behaviour is normally understood via performance than via the mathematics.

Mesh layouts
The principles for mesh layout as discussed for plane stress situations in Section 4.3 are also relevant to plate bending elements.

6.2.2   Validation information for biaxial plate bending

- *Linear elastic behaviour* – see Section 7.2.
- *Normals remain straight* – this assumption tends to be satisfactory except when the span-to-depth ratio is low. It tends to be valid for conventional structural span-to-depth ratios.
- *Shear deformation* – with span-to-depth ratios greater than 10:1 it is very unlikely that shear deformation will be important. As with shear deformation in beam elements, the shear deformation is likely to affect the displacements more than the stresses. If there is doubt as to the validity of neglecting shear deformation, a 'thick plate' (e.g. a *Mindlin*) element can be used. A simplified check on the effect of shear deformation on displacement can be made using an equivalent beam to represent the behaviour and applying equation (5.6).
- *Small deflections*, i.e. that the radius of curvature is equal to $\partial^2 w / \partial x^2$ – as for equation (5.2) for uniaxial bending. This is satisfied if the deflections are small in comparison with the plate dimensions. For thin plates, a reasonable limit on this may be that the maximum deflection should not be greater than the plate depth. Normal deflection criteria in structural design tend to ensure that the small deflection assumption is valid.
- *In-plane actions ignored* – there is no strain in the centroidal plane of the plate. This assumption can be removed by adding in-plane actions to the bending actions, as used for flat shell elements (Section 4.2.3).

6.2.3   Output stresses and moments
Finite element programs normally provide output of stresses, internal moments and shears for plate bending elements.

The traditional notation (Timoshenko and Woinowsky-Krieger 1964) for internal moments and shears in plates is as shown in Fig. 6.2. The moments are explained here in relation to the sign convention of Fig. 6.2, but in finite element software output the signs of the output may not follow this convention.

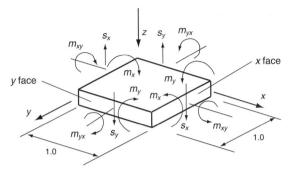

Figure 6.2 Notation for plate bending.

Note the following features.

- All moments and shears are defined per unit width of plate.
- An 'x face' is defined as the face which is normal to the x axis. The 'positive x face' is the one from which the x axis points outwards. The notation for the internal moments are defined in relation to the faces on which they act rather than in relation to the axis about which they rotate.
- m_x and m_y are the direct moments per unit width, and are obtained by integrating the moments of the forces due to the σ_x and σ_y direct stresses, respectively. They are defined as sagging positive (in this case).
- m_{xy} and m_{yx} are the twisting moments per unit width, and are obtained by integrating the shear forces due to τ_{xy} and τ_{yx} stresses, respectively (note that, for example, τ_{xy} is the shear stress on the x plane in the y direction). The twisting moments are characterised as m_{xy} since $|m_{xy}| = |m_{yx}|$.
- m_x, m_y and m_{xy} form a 'triad' which is used to calculate principal moments, etc.
- s_x and s_y are the transverse shear forces per unit width on the x and y faces, respectively. They are the integrals of the τ_{xz} and τ_{yz} stresses, respectively.

Relationships between the internal force actions and stresses for a flat plate are given in Table 6.1.

Table 6.1 Stress components for flat plate bending

Stress component	Maximum value	Position of maximum within the depth of the plate	Shape of stress block
Direct stress	$\sigma_x = 6m_x/t^2$ $\sigma_y = 6m_y/t^2$	Top and bottom	Linear
Shear stress due to twisting moments	$\tau_{xy} = 6m_{xy}/t^2$	Top and bottom	Linear
Shear stress due to transverse shear	$\tau_{xz} = 1.5s_x/t$	Centre	Parabolic

Notes: t – plate thickness.

6.2.4   Checking models for plates in bending

For formal solutions to plates with various boundary conditions and loading see
Young and Budynas (2001) and Timoshenko and Woinowsky-Krieger (1964).

A common assumption is that the slabs span one way. A strip of unit width
spanning in the direction of the shorter span (it will tend to span in that direction)
should be selected.

6.3   Concrete slabs

6.3.1   General

Traditional approaches to defining the internal forces in concrete slabs for
structural design include:

- assuming that the slab spans only in one direction
- using code of practice rules for slab design – for example, flat slab moments are
  assigned to column strips and middle strips using empirical rules
- assuming equivalent widths of slab to act compositely with the support beam.

The modern alternative is to use either a 3D elastic plate element model or a plate
grillage model (Section 6.3.6). Such models are now quite easy to establish, but the
consequent calculations for reinforcement requirements (Section 6.3.3) are more
complex than the traditional methods. It is best to have software to do this
processing.

Concrete slabs can also be analysed using the yield line method, which is briefly
described in Section 6.3.8.

Advantages of using elastic element models for the design of concrete slabs
include:

- no details of the reinforcement are needed to do the analysis
- the process takes some account of serviceability (in that reinforcement is
  specified in relation to the working load situation rather than the failure
  condition), whereas the yield line method does not
- non-regular geometry can be taken into account without special difficulty.

6.3.2   Element models for slab analysis

For element modelling of concrete slabs, the following options are available.

- *Plate bending elements* (Section 6.2) – these elements can be used to model slabs
  where there are no resultant forces in the plane of the slab. Their use tends to be
  superseded by the use of flat shell elements, which do not have this limitation.
- *Flat shell elements* (normally thin shell elements) – incorporate plate bending
  and plane stress components (Section 4.2.3) and can be used in a wide range
  of situations.
- *Plate grillage model* – the grillage model of a plate treats the slab as an inter-
  connection of grillage elements, as described in Section 6.3.6.

In general, if the slab has a mainly flat soffit then plate bending or shell element
models tend to be preferable. Where there are a significant number of beams
and/or ribs, the plate grillage model may be more convenient.

6.3.3   Reinforcing moments and forces for concrete slabs

It is essential to take account of the twisting moment when calculating reinforcement areas for bending of concrete slabs. The Wood–Armer rules are normally used for this (Wood 1969, O'Brien and Keogh 1999). The Wood–Armer output gives reinforcing moments for the top and bottom of the slab, and a conservative approach is to ensure that the provided reinforcement is always greater than that required via the rules (Ades and MacLeod 1992). The Wood–Armer rules should also be used for plate grillage models where the elements of the grillage model include torsional stiffness.

Reinforcement for resultant forces in the plane of the slab (when using flat shell elements) can be based on Clark's rules (Clark 1976).

6.3.4   Plate bending and shell element models
Flat slabs
For information about modelling flat slabs in buildings, see Whittle (1985).

Beam supported slabs – basic modelling principles
It is normal in the structural design of a composite concrete floor slab to assume that it spans in one direction – in the short span of the panel. In the long span direction, the slab is composite with the steel beams. This gives a lower bound model (Section 2.3.3).

The real behaviour of a slab supported on beams will depend on the length-to-breadth ratio of the slab and the ratio of slab stiffness to support beam stiffness. For slabs on rigid supports, a one-way spanning assumption tends to become realistic for slabs with length-to-breadth ratios greater than 2.0. With beam supported slabs, even if the length-to-breadth ratio is greater than 2.0, the moments for spanning in the long direction may be greater than those for the short direction if the two-way action is properly modelled.

Modelling the effect of eccentricity of beams in relation to the slab centreline
Composite steel beam and slab construction
To model the eccentricity between the plane of the slab and the plane of bending of the beams, a rigid link is inserted as shown in Fig. 6.3(a). The rationale for having a rigid link of this kind is discussed in Section 5.6.2.

Without the rigid link there is no composite action between the slab and the beam. With the rigid link there are resultant internal actions in the plane of the slab and therefore a flat shell element is needed for the slab – see the following explanation.

It should be noted that modelling the composite action using rigid links does not require the definition of a centroidal axis for a composite section. The interaction between the beam and the slab is automatically incorporated by the element model.

Implementing the rigid link
It is common to be able to specify an eccentricity at the nodes e (Fig. 6.3(a)) as data for shell elements or for beam elements. It is important to ensure that the correct sign for the eccentricity is used. This is the easiest way to insert a rigid link, but a

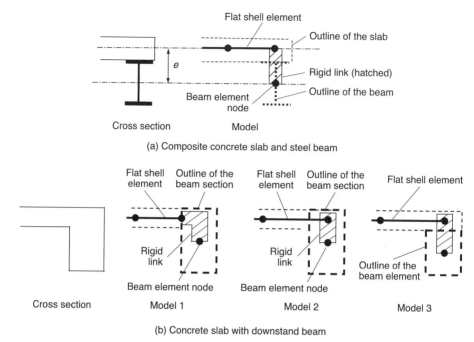

Figure 6.3 Modelling of slabs with beam supports.

constraint equation or an element with high but finite stiffness can also be used (Section 4.5.2). The eccentricity can be defined for the beam element or for the slab element, but not for both.

Concrete downstand beam

Figure 6.3(b) shows three models for an integral downstand concrete beam. In the model 1 case, the boundary between the shell element mesh and the beam is at the edge of the beam (rather than at the centre line of the beam, as in model 2). This requires the rigid link to have a horizontal and a vertical offset. Elements having eccentricities other than in the local z direction are less common and therefore a constraint equation or an element with high but finite stiffness can be used.

If model 2 is used (i.e. the horizontal component of the rigid link is ignored), the shell element and the beam element intersect and the top part of the beam would be included twice in the model – which is not recommended.

It tends to be easier to implement a rigid link using a local z direction eccentricity as element data, in which case model 3 is a possibility. The drawback to this model is that the beam is defined in two parts, making decisions about design axial forces and bending moments for the beam difficult.

Explanation for the resultant axial forces in the slab with composite beams

Figure 6.4(a) shows a steel beam acting compositely with a portion of a concrete slab. Figure 6.4(b) shows the stress block for the equivalent beam in steel, where

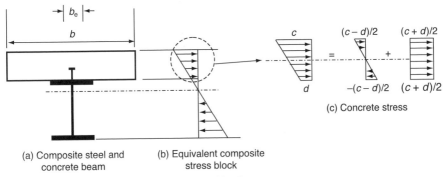

(a) Composite steel and concrete beam

(b) Equivalent composite stress block

(c) Concrete stress

Figure 6.4 The components of a stress block.

the concrete is treated as having an effective width b_e

$$b_e = b/\eta \tag{6.1}$$

where b is the width of the concrete considered to act as a flange to the beam and η is the modular ratio E_s/E_c.

The stresses shown in Fig. 6.4(b) are the stresses in the equivalent steel section. If the actual concrete stresses were shown, the overall stress block for the composite section would be discontinuous at the steel–concrete interface.

The part of the stress block of Fig. 6.4(b) that relates to the concrete is shown in Fig. 6.4(c). The concrete stress block has a trapezoidal shape, which can be divided into a bending component and an axial component, as shown in Fig. 6.4(c) (note that compressive stress is treated as positive for convenience in this context).

The shell elements provide two main components, namely:

- *bending*, which models the bending component of the stress block – the $(c-d)/2$ component of Fig. 6.4(c)
- *plane stress*, which models the uniform stress component of the stress block – the $(c+d)/2$ component of Fig. 6.4(c).

The stress block for the steel beam also has a trapezoidal shape, which is treated as bending and axial components by the beam element.

6.3.5 Shear lag effect

Figure 6.5(a) shows a slab with beam supports along the longer spans, with the shorter spans rigidly supported. Figure 6.5(b) shows the cross section. The slab takes uniformly distributed loading, as shown in the cross section. An analysis model was set up, with the slab treated as a 12×6 mesh of shell elements and the supporting beams as eccentric (as in Fig. 6.3(a)) engineering beam elements. The model results gave the axial force per metre in the slab at Section A–A, as shown in Fig. 6.5(c). Note how the axial force is maximum at the beams and minimum at the centre of the slab. This is called *shear lag*. It is the reason why an equivalent width of slab is used for the flange of the composite beam rather than half the span.

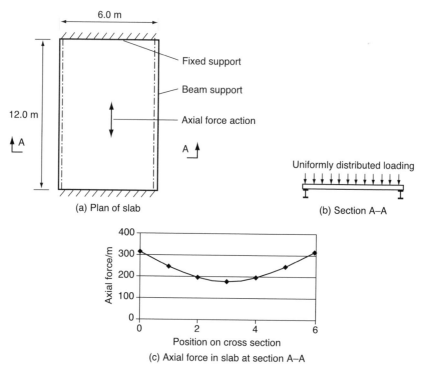

Figure 6.5 *Shear lag effect in a slab.*

6.3.6 *Plate grillage models for concrete slabs*

For a plate grillage model, the slab is divided into strips that are treated as grillage or 3D beam elements (Fig. 6.6). The properties of the elements are based on the cross sections of the strips.

O'Brien and Keogh (1999) and Hambly (1991) give good information for grillage models of bridge decks. Only a brief description is provided here.

Modelling issues

- *Element properties* – Table 6.2 gives typical element properties. The second moment of area and shear area of the element is that of the normal cross section, but note that for a slab element of uniform depth the torsional constant J is $bd^3/6$, rather than $bd^3/3$ as for a separate strip of material. The difference is due to the lack of vertical shear flow at the edges of a grillage element due to its continuity with adjacent elements (Hambly 1991).
- The m_x, m_y and m_{xy} moments in an element need to be converted to reinforcing moments using the Wood–Armer or other rules (Section 6.3.3). For this purpose it is important to ensure that it is the moment per unit width that is used (moment from the grillage element divided by the width of the element).
- The centroidal axis of an element does not need to coincide with the centroid of the area (Whittle 1985).

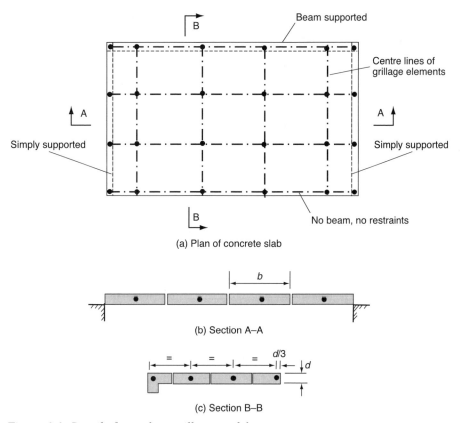

(a) Plan of concrete slab

(b) Section A–A

(c) Section B–B

Figure 6.6 Details for a plate grillage model.

- For *edge elements* the element axis can be at a distance $d/3$ from the edge, where d is the slab depth (Hambly 1991) (see Fig. 6.6(c)).
- *Uniform transverse loading* can be distributed to the grillage members on the basis of the conventional trapezoidal distribution for a rectangular slab, but Whittle (1985) suggests that uniformly distributed loads on the grillage members based on a distribution in proportion to the panel dimensions gives adequate accuracy. That is, if the side lengths of a panel taking a total

Table 6.2 Section properties for grillage models

Section	I	J	A_s
$b \times d$	$bd^3/12$	$bd^3/6$	$5bd/6$
b, d, d_s, d_w, b_w	As for symmetric bending	$bd_s^3/6 + \dfrac{3b_w^3 d_w^3/10}{b_w^2 + d_w^2}$	$0.85 b_w d$

uniformly distributed load W are L_1 and L_2 then the adjacent beams take a uniformly distributed load of $W/2/(L_1 + L_2)$ kN/m.

- Allowance can be made for *cracking* to estimate service load deflection (see Whittle 1985).
- With *ribbed* or *beam and slab* systems the choice of elements takes account of the variations in cross section. For example, a support beam with associated slab flange can be treated as one of the grillage elements, as illustrated in Fig. 6.6(c). A rib with associated slab can be treated as an element, or a group of ribs can be so treated (Section 6.3.7).

Validation information for the plate grillage model

- The grillage model only roughly approximates to the thin plate biaxial plate bending model (Section 6.2). It tends not to model torsional effects accurately.
- A 'lower bound' solution is obtained – see Section 2.3.3.
- As the mesh density is increased, the solution will tend to converge towards a limit which may be close to, but not the same as, the biaxial bending 'exact' solution. Gordon and May (2004) reported that in some cases, refinement of grillage meshes does not give converging results and that the plate grillage model can give results that are significantly less accurate than using plate bending elements. In particular, the prediction of twisting moments can be of low accuracy, hence the plate grillage modelling of skew slabs may give poor results.
- Use of the plate grillage model may be advantageous where the system is grillage-like – for example in a building where there is a transfer structure of beams to cater for a change in column layout at a floor level.
- A main advantage of the grillage model is that it produces stress resultants that are defined over a specified width and so can be used directly to calculate reinforcement requirements over that width. Plate bending models tend to be more accurate, but they give values of stress resultant per unit width and so the user has to decide over what widths to integrate these for reinforcement calculations. The use of a good postprocessor for plate bending element models is recommended.

6.3.7   Ribbed slabs
Plate grillage model
A slab with equal ribs in both directions can be modelled as a plate grillage (Section 6.3.6), where a single rib with associated slab can be treated as a grillage element or a set of ribs can be modelled as a grillage element (Fig. 6.7(a)). If the rib geometry is different in the orthogonal directions then the appropriate rib stiffnesses are defined for the relevant directions. This is known as *geometrically orthotropic*.

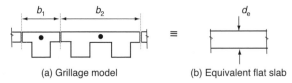

(a) Grillage model   (b) Equivalent flat slab

Figure 6.7   Grillage model for a ribbed slab.

Flat plate model
Geometrically isotropic slabs
A ribbed slab can be modelled using flat plate elements with an equivalent thickness, based on

$$I_{ep} = I_{rs} \tag{6.2}$$

where I_{ep} is the I value per unit width of the equivalent plate and I_{rs} is the I value per unit width of ribbed slab.

For example, if the I value for the single rib element with width of flange b_1 shown in Fig. 6.7(a) is I_{rs} and the equivalent slab thickness is d_e (Fig. 6.7(b)), then equation (6.2) becomes

$$d_e^3/12 = I_{rs}/b_1$$

hence

$$d_e = \sqrt[3]{12I_{rs}/b_1} \tag{6.3}$$

Equation (6.3) is for slabs that are geometrically isotropic (ribs are the same in both directions).

Geometrically orthotropic slabs
Slabs that have different ribs at right angles (geometrically orthotropic) can be modelled as being materially orthotropic (O'Brien and Keogh 1999) by modifying the E_x and E_y values (Section 7.2.3) such that

$$E_{slab}I_{x,slab} = E_{x,plate}I_{x,plate} \tag{6.4}$$

and

$$E_{slab}I_{y,slab} = E_{y,plate}I_{y,plate} \tag{6.5}$$

where 'slab' refers to the actual slab and 'plate' refers to the finite element model, and the I values are per unit width.

This leads to the following rules for defining the element properties.

1. Calculate $I_{x,slab}$ and $I_{y,slab}$.
2. Plate thickness $d_e = \sqrt[3]{12I_{x,slab}}$.
3. $E_{x,plate} = E_{slab}$.
4. $E_{y,plate} = E_{slab}I_{y,slab}/(d_e^3/12)$.
5. $\nu_{xy} = \nu_{yx} = \nu_{slab}$.

The values of the torsional constant G_{xy} can be estimated by substituting the values of E_x and E_y into equation (7.8).

Ribbed slab elements
Some finite element packages have a special plate bending element for ribbed slabs.

6.3.8 Plastic collapse of concrete slabs – the yield line method
It is possible to estimate the plastic collapse load for a slab by defining *yield lines* and identifying the pattern of lines that gives the lowest prediction of collapse load

Yield line

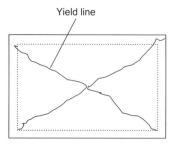

Figure 6.8 Plan of a simply supported slab showing yield lines at failure.

(upper bound approach). A yield line is a continuous plastic hinge in the slab. Figure 6.8 shows yield lines for the collapse of a uniformly loaded concrete slab simply supported on four sides.

For information about the yield line method see Moy (1996).

Validation information

- The yield line method can be useful for checking the ultimate load of an existing slab if there is doubt about its strength.
- The yield line method can give reasonable predictions of collapse load of a slab with statically determinate supports. However, with statically non-determinate supports (as in Fig. 6.8), secondary membrane action (i.e. due to in-plane forces developing in the plane of the slab due to finite geometry of the deformation) may result in real collapse loads that are much greater than predicted by the yield line method.
- If the yield line approach is used to design reinforcement sizes then it is especially important to check for deflection and cracking because the method takes no account of serviceability.

7 Material models

7.1 Introduction

The understanding of material behaviour is a major issue in modelling. This chapter gives a basic introduction to material models.

7.2 Linear elastic behaviour

7.2.1 General

The term *elastic* implies that when the material is unloaded it follows the same deformation path as when loaded. *Linear* implies that stress and strain (and hence load and displacement) vary proportionately (Fig. 7.1).

The word 'elastic' and its derivatives (e.g. 'elasticity') tend to be used with the meaning 'linear elastic', but some materials (e.g. rubber) have non-linear elastic properties.

Most materials used in structural engineering can be modelled as linear elastic up to limiting stress levels. The main advantages of this material model are that solutions can be easily and rapidly produced, and the principle of superposition (Section 2.3.2) is valid. Non-linear material models tend to be more complex and with much longer solution times. The main disadvantages of elastic models are that they cannot predict deformation beyond the elastic limit and can only predict failure of brittle materials.

An elastic material is normally assumed to be *continuous*, i.e. its properties are smoothly distributed over its volume. Whether or not a material can be described as continuous depends on the viewpoint. Masonry does not have the appearance of a continuous material but concrete does. However, if a cut section of concrete is examined, the high degree of discontinuity in the structure is evident. Therefore,

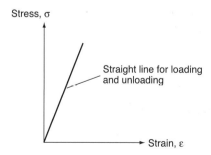

Figure 7.1 Linear elastic behaviour.

a large piece of masonry may be no less discontinuous than a piece of concrete. Steel is a discontinuous material at a micro level.

7.2.2   Types of elastic behaviour

Isotropic is the most common form of elasticity. This assumes that the stiffness of the material is the same when measured in any direction. The elastic behaviour of an isotropic linear elastic material can be defined by two parameters – Young's modulus E and Poisson's ratio ν.

- *Young's modulus E* is the ratio between uniaxial direct stress increment $d\sigma$ and the corresponding direct strain increment $d\varepsilon$ in a tensile test.

$$E = d\sigma/d\varepsilon \qquad (7.1)$$

  or if there is no initial stress or strain

$$E = \sigma/\varepsilon \qquad (7.2)$$

  where σ is the direct stress and ε is the direct strain in the material.

- *Poisson's ratio ν* for an applied strain in the x direction is

$$\nu = \frac{\text{strain increment at right angles to the applied strain}}{\text{applied strain increment}} = \frac{d\varepsilon_y}{d\varepsilon_x} \qquad (7.3)$$

  or if there is no initial stress or strain

$$\nu = \varepsilon_y/\varepsilon_x \qquad (7.4)$$

  where ε_x is the direct strain in the x direction and ε_y is the direct strain in the y direction.

The *shear modulus G* is the ratio of shear stress τ to shear strain γ. For an elastic material with no initial stress or strain.

$$G = \tau/\gamma \qquad (7.5)$$

where τ is the shear stress and γ is the shear strain.

For an elastic isotropic material, the value of G is not normally required to be input since it is related to E and ν by the relationship

$$G = E/(2(1+\nu)) \qquad (7.6)$$

Anisotropic behaviour is where the stiffness of the material is not the same in all directions.

Orthotropic behaviour is where the material has different stiffnesses in two directions at right angles, requiring two E values to define its behaviour – for example timber, which has different properties with the grain and at right angles to the grain.

7.2.3   Values of elastic constants
Isotropic materials

Table A6 gives typical values E and ν for some isotropic structural materials. It is uncommon for the behaviour of a structural system to be sensitive to the value of Poisson's ratio and therefore the values for ν quoted in Table A6 will normally be adequate.

Orthotropic materials
The normal input parameters for an orthotropic material are E_x, E_y, ν_{xy} and G_{xy}, where E_x is Young's modulus in the x direction, E_y is Young's modulus in the y direction, ν_{xy} is Poisson's ratio for loading in the x direction and G_{xy} is the shear modulus for orthotropic conditions.

The values of E_x, E_y and ν_{xy} can be established from measurements of tensile test specimens cut from a sample of the material in the principal directions of orthotropy.

Note that for an orthotropic material

$$\nu_{yx}/E_y = \nu_{xy}/E_x \tag{7.7}$$

hence only ν_{yx} or ν_{xy}, but not both, needs to be input.

The value of the shear modulus for orthotropic materials is problematical. The following empirical relationship is sometimes used (Troitsky 1967).

$$G_{xy} = \frac{\left(1 - \sqrt{\nu_{xy}\nu_{yx}}\right)\sqrt{E_x E_y}}{2\left(1 - \nu_{xy}\nu_{yx}\right)} \tag{7.8}$$

7.2.4   Validation information for linear elastic materials

- *Yield criterion not violated* – the elastic condition may be validated by imposing the requirement that a yield criterion, such as the Von Mises criterion, must not be violated within the system for any load condition.
- *Design to code of practice* – the sizing of members to a code of practice tends to result in stresses at working load which are below the elastic limit. Therefore, the criterion "Design to code of practice number *xxx*" may be acceptable.
- *Lower bound theorem* – elastic analysis is commonly used to define a set of internal forces that are then used to size the member for ultimate load. This may be justified on the basis of the *lower bound theorem* of plasticity (Section 2.3.3).
- *Steel* – linear elasticity is a good model for steel up to yield.
- *Concrete* – a criterion sometimes used for elastic behaviour in concrete is $f_e < f_c/3$, where f_e is the limiting stress in the concrete for elastic behaviour and f_c is the cube or cylinder strength of the concrete. Concrete is brittle in tension, but it can, to a degree, behave plastically in compression. To predict short-term deformation at low stress levels, an elastic model may be acceptable.
- *Long-term movements* – concrete is susceptible to significant long-term deformation due to creep (i.e. deformation due to sustained load) and shrinkage (i.e. compressive deformation due to loss of water from the concrete). It was formerly common practice to use a factorised value of the E value for concrete to predict long-term behaviour. A typical criterion for linear elasticity for concrete is to divide the short-term E value by 2.0 to account for long-term effects. This can give a significant underestimate of the long-term movement. The long-term deformation of concrete has little relationship with the E value. Long-term movement of concrete can continue over many years, and deformations can exceed the short-term elastic values by

a factor much greater than 2.0. Codes of practice for structural concrete have provisions to allow for creep and shrinkage. For more information on this topic see Ghali and Favre (2002).

- *Timber* – the linear elastic model tends to be satisfactory for timber up to certain stress levels. The stiffness of unlaminated timber is strongly ortho-tropic, with the stiffness at right angles to the grain being less than that parallel to the grain. It is normally only loaded in the direction parallel to the grain and therefore modelled in the uniaxial condition. Laminated and composite timber sheets may be treated as being isotropic.

- *Brick masonry* – brickwork behaves in a similar way to concrete, with low tensile strength and being brittle in tension. It also suffers from creep in a similar way to concrete, but is possibly less susceptible to creep than concrete. Whereas new concrete tends to shrink due to loss of moisture, new brickwork tends to expand due to take up of moisture (this is because bricks coming from the kiln have a very low water content).

7.3 Non-linear material behaviour

7.3.1 Plasticity

Uniaxial stress

The term *plastic* implies that when the material is loaded and then unloaded, the deformation paths are not the same and there will be residual deformation.

A *brittle* material fails without any plastic deformation.

Plasticity is a condition in which non-recoverable deformation can take place. Most structural materials behave elastically up to a limit and then fail (brittle material) or have a plastic phase prior to failure. Failure is when the material is no longer capable of taking load.

When using a plasticity assumption, the fact that no structural material has unlimited ductility must be taken into account.

The *yield point* of a material marks the onset of plasticity, i.e. where some of the deformation starts to become non-recoverable. With uniaxial stress (e.g. in the tensile test), the point of yield may not be easy to define precisely as there may not be a sharp transition between elastic and plastic zones. However, a yield stress is defined as σ_Y, marking the onset of plasticity (Fig. 7.2).

An important issue in relation to plasticity is ductility. This is the deformation capacity beyond yield, and is normally defined by a ductility factor μ, where

$$\mu = \frac{\text{strain at ultimate stress}}{\text{strain at yield stress}} = \frac{\varepsilon_2}{\varepsilon_1}$$

or

$$\mu = \frac{\text{displacement at ultimate load}}{\text{displacement at yield load}} \tag{7.9}$$

where ε_1 is the strain at first yield and ε_2 is the strain at failure (see Fig. 7.2).

A common material model for plasticity is *elastic-perfectly plastic*, where beyond the plastic limit the material continues to withstand its yield stress σ_Y but does not take further stress (Fig. 7.3(a)). If unloading takes place from beyond the σ_Y level it

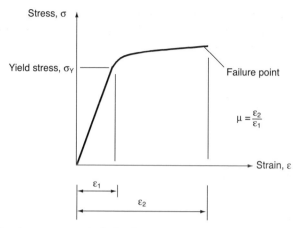

Figure 7.2 Plastic stress–strain behaviour.

is assumed that the stress–strain relationship during unloading is linear and parallel to the elastic loading curve (Fig. 7.3(b)). At zero load after unloading, there will be a residual plastic strain ε_p.

Mild steel behaves fairly closely to this ideal. For most other ductile materials, it is a fairly rough approximation.

Another common non-linear material relationship is *elastic-plastic with strain hardening* (Fig. 7.3(b)). This derives its name from the fact that after the material has had a cycle of loading beyond the yield stress, in a later cycle of loading the onset of plasticity is at a higher level than in the first cycle.

Biaxial stress

With biaxial stress the onset of plastic behaviour can be defined by a number of yield criteria. For ductile materials, the Tresca and the Von Mises yield criteria are used. The Tresca condition is based on maximum shear stress and is more conservative than Von Mises, which is based on maximum shear strain energy.

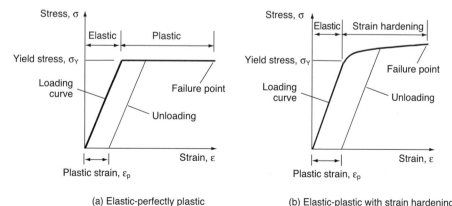

(a) Elastic-perfectly plastic   (b) Elastic-plastic with strain hardening

Figure 7.3 Basic stress–strain models.

Associated with plastic behaviour is a 'flow rule', which defines the post-yield deformation characteristics. A common approach is to model the post-yield stress–strain relationship as a series of straight lines.

7.3.2   Other non-linear constitutive relationships

Major finite element packages make provision for a wide range of non-linear material conditions, but discussion of these is outside the scope of this text.

8 Support models

8.1 Introduction

An analysis model of a structure has to be defined in relation to some frame of reference; it has to be fixed in space; it has to be supported.

In this context the *structure* includes the superstructure and the foundations, and *ground* is what is below the foundations, including soil and rock. Four basic ways of defining the supports for a structure are:

- *Support fixity model* (Fig. 8.1(a)) – deformations of the ground are ignored and the nodes for the structure at the contact with the ground are given fixed restraints.
- *Winkler model* (Fig. 8.1(b)) – the ground is modelled by linear elastic springs at the structure–soil interface. The springs are not coupled, i.e. when one spring deforms the other springs are unaffected by shear transfer in the ground.
- *Half space model* (Fig. 8.1(c)) – the ground is modelled by coupled springs at the structure–soil interface, i.e. shear transfer in the ground is considered.
- *Element model for the ground* (Fig. 8.1(d)) – the ground is modelled as finite elements which have fixities at distances from the structure.

These four types of support model are discussed in this chapter.

8.2 Modelling support fixity

8.2.1 General

Restraint

A *restraint* is a limit placed on the movement corresponding to a specific degree of freedom (defined in Section 4.2.6) of the model in relation to the reference system.

For a *fixed restraint* the movement is set at zero, whereas a *spring restraint* (Section 8.3.2) has a (normally) linear elastic spring connection to the fixed reference system.

Support

The term *support* normally refers to a node that has restraints. Table 8.1 shows conventions for indicating supports in analysis models. For information about modelling spring supports, see Section 8.3.2.

8.2.2 Support requirements

To each restrained freedom of the structure there corresponds a reaction force. These reaction forces must at least form a set which is statically determinate.

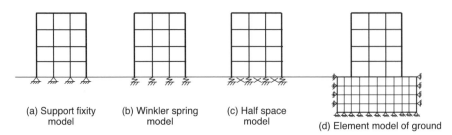

Figure 8.1 Models for structure support.

Equation solvers will fail if the supports are not sufficient to fix the system in space.

If a structure such as an aircraft in space is to be modelled for static forces then it should have a set of determinate supports and the applied loading should be self equilibrating. There should then be no reactions at the restraints, and the deformations will be relative to the restrained directions. For dynamic analysis of structures in space this process is not valid, but some eigenvalue extraction routines can deal with the situation (NAFEMS 1992).

Neglecting fixity where there is a degree of restraint will tend to be conservative for estimates of deformation and internal forces. It may be justified by the lower bound theorem (Section 2.3.3).

8.2.3 Roller supports
When a beam is supported on a roller bearing then the support in the model is treated as a roller. When the supports for a beam are not designed to roll it is still normal to neglect any horizontal restraint. Figure 8.2 shows conventional

Table 8.1 Symbols for restraints

Symbol	Type of restraint	Restrained freedoms
	Horizontal roller	
	Pin	
	Fixed	
	Vertical roller – no rotation (at axis of symmetry)	
	Spring	Translational
	Spring	Rotational

(a) Simply supported beam   (b) Continuous beam

Figure 8.2 Beam support modelling conventions.

models for single and continuous beams. They are assigned one pin support – to give a single horizontal restraint – with the other supports as rollers. (Without at least one horizontal restraint the equation solver would fail, even if there are no horizontal loads.) With only vertical loads on beams and with conventional span-to-depth ratios, the choice of roller or pin supports will tend to be unimportant (but not if temperature movement is included in the model).

Roller supports for deep beams

In a frame model, the restraints are normally specified at nodes defined at the centroidal axes of members. However, for a beam sitting on a wall (for example) the real point of support is normally at the bottom of the beam rather than at the centroidal axis (Fig. 8.3(a)). Figure 8.3(b) shows how supports are normally applied at the centroidal axis for such beams. Figure 8.3(c) shows a more accurate representation, where the supports are defined at the bottom of the beam by specifying rigid offsets from the centroidal axes (see Section 5.6.2).

If the beam is simply supported (i.e. it has a statically determinate set of supports reactions) as shown in Figs 8.3(b) and (c) then the internal force actions (due to transverse loading as shown) will not be affected by the presence or absence of the rigid offsets. However, if there is horizontal restraint at both of the supports then horizontal forces will be induced in the support system by the transverse load. The following case study demonstrates a situation where this action was not negligible.

Case study – roller support for a deep beam

Brick beams were being tested to failure to assess the effectiveness of bed-joint reinforcement for in-plane bending (MacLeod 1987). The set-up was as shown in Fig. 8.4(a). Just before failure a crack had formed from the bottom of the specimen to near the top, suggesting that the depth of the compression stress block was about the depth of one course of brickwork (Fig. 8.4(c)). To calculate the stress in the steel, a free body diagram, as in Fig. 8.4(b) but neglecting the horizontal force H (i.e. assuming no horizontal restraint at the support), was used. Taking moments about the centre of compression – at point a on Fig. 8.4(b) – indicated a tensile failure stress in the steel reinforcement of over $700\,\mathrm{N/mm^2}$, as compared with a tensile test measurement value of the order of

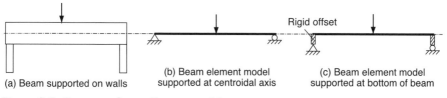

(a) Beam supported on walls

(b) Beam element model supported at centroidal axis

Rigid offset

(c) Beam element model supported at bottom of beam

Figure 8.3 Supports for wall-supported beam.

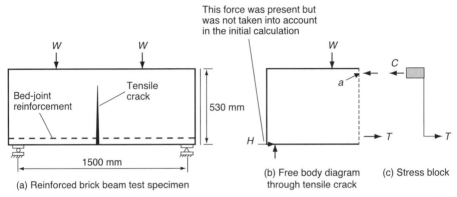

(a) Reinforced brick beam test specimen

(b) Free body diagram through tensile crack

(c) Stress block

Figure 8.4 Deep brick beam test.

$520\,\text{N/mm}^2$. After much puzzling it was realised that the free body diagram was wrong; there was a horizontal force at the supports – the roller (a steel pin between two plates) was not rolling freely. Devising one that did roll freely proved to be quite difficult (it was necessary to find a bearing with a stiff shell).

8.2.4 Pin supports
As with roller supports, even if the support is a real pin (Fig. 5.16) then friction will tend to induce partial restraint. The assumption of a pin where there is clearly some rotational restraint may be justified on the basis of being conservative or on the basis of the lower bound theorem (Section 2.3.3).

8.2.5 Rotational restraint at a cantilever support
Rotational fixity at a cantilever support will be realistic for the prediction of internal forces since the cantilever is statically determinate, but it may not be acceptable in relation to deformation. Figure 8.5 shows the deformations of a cantilever. The tip deflection is due to bending along the length of the member, Δ_1, and to the rotation at the support, Δ_2. The tip deflection Δ of the cantilever shown in Fig. 8.5 will be (assuming small deformations and neglecting shear deformation):

$$\Delta = \Delta_1 + \Delta_2 = \frac{WL^3}{3EI} + L\theta \tag{8.1}$$

where θ is the rotation at the support. The $L\theta$ term in equation (8.1) acts as a magnifier on the tip deflection of the cantilever. It is often significant and may be dominant.

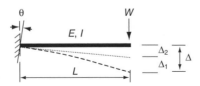

Figure 8.5 Cantilever with support rotation.

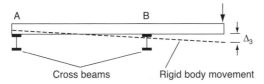

Figure 8.6 *Beam supported on cross beams.*

If a linear spring is defined to model the rotational behaviour at the support then θ may be calculated using

$$\theta = M/K_\theta \tag{8.2}$$

where M is the moment at the support and K_θ is the rotational stiffness of the support (kN m/rad).

Equation (8.1) (or equivalent relationships for other loading conditions) can be used to assess the effect of the support rotation on the end deflection of a cantilever.

Figure 8.6 shows a beam with a cantilever end supported on two cross beams. In such a situation, three components will contribute to the tip deflection of the cantilever:

- the bending of the cantilever over its length, Δ_1
- the support rotation at B due to bending, Δ_2
- under the loading shown in the diagram, the cross beams will move up at A and down at B resulting in a rigid body displacement of the main beam as shown – this will add a third component to the tip deflection, Δ_3.

8.2.6 Rotational restraints at column bases

The assumption that a column is fully restrained at its base may result in an overestimate of stiffness and an underestimate of the maximum frame moments. Issues to be considered when validating a fixed column support include:

- whether or not the stiffness of the frame is a critical issue
- the detailing at the support – is the foundation sufficiently massive that the rotational stiffness at the support will be negligible?
- the lower bound approach (Section 2.3.3) may be invoked if the frame is designed to take the moments from the analysis.

BS 5950 (BSI 2000) requires the following rules for modelling the rotational restraint to a steel column.

- Column rigidly connected to a foundation:
  base rotational stiffness $= 4(EI)_c/h$
- Column nominally connected to the foundation:
  base rotational stiffness $= 0.1(4(EI)_c/h)$

where $(EI)_c$ is the bending stiffness parameter of the column and h is the height of the column.

Elastic spring supports can be modelled as described in Section 8.3.2.

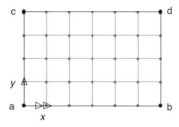

Figure 8.7 Simply supported mesh of plate bending elements.

Table 8.2 Restraints for a simply supported slab

Line	Δ_z	θ_x	θ_y
Lines ab, cd	R	F	R
Lines ac, bd	R	R	F
Corner nodes	R	R	R

Notes: Δ_z – displacements in the z direction; θ_x, θ_y – rotation about the x, y axes; F – free; R – restrained.

8.2.7   Slab supports

The principles for beam supports also apply to slab systems. The assumption of fixed supports is likely to give economies in the design for moment but will tend to underestimate deflection. Figure 8.7 shows a plate bending element model of a slab with all sides simply supported. The restraints required for this condition are defined in Table 8.2.

8.3   Modelling the ground

8.3.1   General

The two main types of 'ground' are soil and rock. The model of a structure is more likely to be realistic if the deformation of a rock support is neglected than if the effect of a soil support is neglected. An elastic model for rock may be realistic in some situations, but the real behaviour of jointed rock is complex.

Soil tends to be non-homogeneous, with mechanical properties that may be time-dependent, functions of water content and non-linear in relation to stress and strain. To address these features requires advanced analysis that is outside the scope of this text.

Taking account of the structure and the ground in a single analysis is known as *soil–structure interaction* (IStructE 1989). This is very difficult to model accurately. For example, one of the main reasons for using a model for a building that combines the structure and the soil is to estimate differential settlement. However, one of the most important factors in relation to differential settlement – the variation of the stiffness of the soil over the footprint of the building – is seldom taken into account in such models (see the 'model refinement principle' discussed in Section 2.2.1).

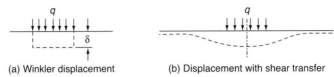

Figure 8.8 Displacement patterns.

It is generally thought that the structure is easy to model relative to the soil, but the real behaviour of a structure depends on the cladding, the internal partitions, temperature effects, etc., which are difficult to represent in a model.

However, because it is very difficult to produce a predictive model does not mean that it is not possible to take account of soil–structure interaction. While 'garbage in, garbage out' needs to be avoided, use of a very approximate model (such as the Winkler model described in the following section) may be better than no attempt to model the interaction.

8.3.2 The Winkler model for soil behaviour
Basis of the Winkler model
Figure 8.8(a) illustrates the deformation of a Winkler support. If a vertical load is applied to a rigid plate, the soil surface is assumed to move downwards with a linear elastic response over the area of the plate but does not deform beyond the plate boundaries. This neglects shear transfer in the soil which makes the deformation extend beyond the limits of the plate, as shown in Fig. 8.8(b). Over the area of the load, the soil acts as a simple spring with the load deformation relationship

$$q = k_{wk}\delta \tag{8.3}$$

where q is the uniform pressure applied to the plate (equal to the reaction pressure of the soil on the plate), k_{wk} is often denoted as the 'modulus of subgrade reaction' but denoted here as the *Winkler stiffness* (it is based on plate bearing tests and has units of kN/m^3), and δ is the deformation of the soil surface.

Validation information for the Winkler model
A 1978 Institution of Structural Engineers' report on structure–soil interaction stated that the Winkler model "cannot be recommended for the analysis of rafts and continuous footings" (IStructE 1978). A later edition of this report (IStructE 1989) does not even mention the model. The model is, however, recommended for use in design of combined footings and mats (ACI 2002).

The main negative feature of the Winkler model is that results from it can be significantly different from those obtained by measurement, i.e. it cannot be treated as a predictive model. In particular, it gives poor predictions of soil pressure at the edges of foundations. However, the method:

- is easy to implement using a structural analysis package
- may give better results than by ignoring the effect of soil stiffness in an analysis of the structure
- may give indicative results, i.e. provide information about trends in behaviour.

Values for k_{wk}

Table A7 gives typical values for k_{wk}. It is possible to relate k_{wk} to the elastic properties of the soil using (Hemsley 1998)

$$k_{wk} = 2E_s/(\pi a(1 - \nu_s^2)) \tag{8.4}$$

where E_s is Young's modulus for the soil (see Section 8.3.4 and Table A8), ν_s is Poisson's ratio for the soil (see Table A9) and a is the radius of the equivalent plate.

Implementing the Winkler model

Most structural analysis software allows elastic spring supports which are, effectively, Winkler springs. In some cases the derivation of the element stiffness matrix includes a Winkler support, and the Winkler stiffness is part of the data for the element.

A single footing can be modelled as a rigid block supported on a Winkler spring with the properties

$$K_v = k_{wk}A \tag{8.5}$$

$$K_\theta = k_{wk}I_f \tag{8.6}$$

where K_v is the vertical spring stiffness, k_{wk} is the Winkler stiffness, A is the area of the footing, K_θ is the rotational spring stiffness and I_f is the I value of the base of the footing (about the same axis as K_θ is defined).

For a 3D model, rotational springs about two axes can be defined.

The value of a Winkler spring stiffness at a node under a raft is calculated using

$$K_{winkler} = k_{wk}A_t \tag{8.7}$$

where k_{wk} is the Winkler stiffness and A_t is the area of the soil surface which is tributary to the node (for example, with a mesh of nodes which has spacing a by b, the tributary area for an internal node is ab, for a side node is $ab/2$ and for a corner node is $ab/4$).

8.3.3 Half space models

The concept of a half space model (Fig. 8.1(c)) is that a grid of nodes is defined at the ground surface and a stiffness matrix for the soil is defined at that level. This is added to the stiffness matrix for the structure to produce a model that incorporates both the soil and the structure. This is also known as a *boundary element* model. Such models are not normally implemented in structural analysis software. Linear elastic and non-linear models can be used.

The *surface element method* (not to be confused with the generic concept in finite element analysis discussed in Section 4.2.3) described by Hemsley (1998) is a potentially useful elastic model for defining a half space model. The stress in the soil is estimated using homogeneous conditions, but the strains caused by these stresses are integrated taking into account variation of properties within the depth of the soil.

8.3.4   Finite element models

To take account of soil–structure interaction, the soil can be treated as an elastic medium and a finite element model used for solution. Because of the power of modern computers this is now a realistic option for soil–structure interaction modelling, especially if elastic properties are utilised.

The following types of elements can be used.

- A 3D model using 3D elements (Section 4.2.4).
- For long structures, a plane strain (Section 4.2.3) model can be used.
- Axisymmetric elements can be used for circular systems with axisymmetric loading.

Using 3D finite elements has the disadvantage that the order of solution may be high, although this may not be significant with modern computing power. Variation in the material properties both horizontally and vertically is easily catered for in a finite element model.

It is necessary to define limits to the extent of the finite soil model. Desai and Abel (1972) quote from other sources the following rules.

$$H/B: 4.0-6.0 \tag{8.8}$$

$$V/B: 10.0-12.0 \tag{8.9}$$

where B is the plan dimension of the structure, H is the horizontal distance from the centre of the structure to the limit of the soil model and V is the vertical distance from the base of the structure to the limit of the soil model (i.e. the depth of the soil model) (Fig. 8.9).

Validation information for modelling the soil as linear elastic

- Significant change in the water content of the soil after loading is likely to invalidate the elastic assumption for predictive analysis.
- In general, the results will tend to be indicative rather than predictive.
- Linear elasticity tends to give better results for cohesive soils than for cohesionless soils.
- Linear elasticity will give better results than the Winkler model.

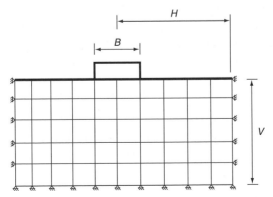

Figure 8.9 Limits of finite model of soil.

- Linear elasticity will tend to give better results for short-term loading than for long-term loading.
- Due to the confining pressure, E values for the soil will increase with depth.

Elastic parameter values
Normally, the undrained values for Young's modulus and Poisson's ratio are used. Tables A8 and A9 give typical values of elastic parameters for soil.

Non-linear material models
Some relevant non-linear constitutive relationships are available in conventional finite element software, for example based on the Mohr–Coulomb criterion (which tends to be suitable for cohesionless soils). However, discussion of the use of such models is outside the scope of this text.

8.4   Foundation structures

8.4.1   Ground beams
Ground beams and continuous footings can be modelled as beam elements.

8.4.2   Raft foundations
Rafts can be treated as plate bending or as plate grillage models, as described in Section 6.3. A cellular raft can be treated as a slab of an equivalent thickness, as described in Section 6.3.7, or can be modelled as a box structure, using either flat shell or volume elements.

Refer to the bibliography for information on the analysis and modelling of rafts.

8.4.3   Piles
The stiffness of a pile would normally be characterised by springs at the top of the pile. The choice of spring stiffness depends on several parameters, including:

- pile geometry
- E value for pile
- soil characteristics
- whether or not the pile is end bearing
- the interaction of piles within a group.

Refer to the bibliography for information on the analysis and modelling of piles.

9 Loading

9.1 Introduction

This chapter discusses validation issues that relate to loading. The term 'loading' in this context implies the general concept of 'external actions on the structure'.

In many cases the validation of the loading can simply refer to a code of practice, but it may be important to think beyond the code. The degree of uncertainty in certain types of loading can represent the greatest source of risk in a modelling context.

9.2 Dead loading

Of the various load types, dead loading is normally associated with the least uncertainty.

9.3 Live loading

Code provisions are normally used for live loading, but it is always worthwhile to assess whether or not the design situation is consistent with the scope of the code being used.

9.4 Wind loading

The introduction of the current UK code for wind loading (BSI 1997) both significantly increased the scope of the provisions and the complexity of their use. Despite this, there may be circumstances where the uncertainty is high and the consequences of failure due to wind are great (e.g. roofs over spaces in which large numbers of people congregate). Where such uncertainty exists, the commissioning of wind tunnel tests may be advisable.

9.5 Earthquake loading

Prior to the computer era, the standard procedure for earthquake design was to use an equivalent lateral static load on an elastic model of a structure – the *quasi-static* approach. A total base shear which depends on the history of earthquakes in the area and other factors is calculated. This shear is then treated as a lateral load with linear variation of intensity with height, from zero at the base to a maximum at the top (Fig. 9.1). This increase in intensity with height is assumed to take account of the 'whipping' action at the top of a structure during a seismic event. This is still the standard method for ordinary structures.

Figure 9.1 Equivalent static load for earthquake analysis.

When computers started to be used for structural analysis, investigations were made with dynamic analysis, where the input spectrum for a known earthquake was applied to an elastic analysis model of a structure. The predicted base shears from this method had values which were significantly greater than were used in the quasi-static method. This led to the conclusion that in an earthquake a building will exhibit post-yield material behaviour. To account for this, the base shears from the dynamic analysis were divided by a 'ductility factor', typically in the range 2 to 4. This factor was, in effect, the ratio between the value you had and the value you wanted to have.

Nowadays, it is common to carry out *modal analysis* for seismic design, where the response for different vibration modes are treated separately and then combined. It can be argued, however, that if elastic analysis is so far from the real behaviour in an earthquake then there is little sense in refining it. In Priestly (2003), several conventional ideas about seismic design, including the ineffectiveness of modal analysis, are seriously challenged.

The modelling of a structure in an earthquake represents one of the most difficult challenges in structural engineering. The input conditions are highly uncertain and the analysis model needs to take account of the real behaviour of the system, including post-yield behaviour, the effect of load reversals, second-order geometry effects and the effects of ground movement. Unless these are taken into account in a realistic way, refinements in the model may deliver illusory benefits. The 'model refinement principle' of Section 2.2.1 needs to be considered.

The major source of information about how structures behave in earthquakes is the observation of structures that have been damaged by earthquakes. Conventional structural analysis does little to promote understanding of behaviour in such real contexts.

It could be suggested that Priestly's ideas about the inadequacy of seismic design procedures (Priestly 2003) casts a shadow of doubt over design procedures for other loading types in that elastic analysis cannot adequately account for ultimate load conditions. An important issue is that for dead load, live load and wind load, while we do design for ultimate conditions, we do not expect the structure in its lifetime to achieve this condition, i.e. we do not expect it to suffer significant post-yield behaviour. However, except for some high-risk structures, such as buildings for nuclear facilities, it is considered to be uneconomic to design a building to remain elastic in a major earthquake. Therefore, in an earthquake, the building is expected to experience complex post-yield behaviour. As a result, post-yield behaviour is much more important for seismic loading than for most other types of loading.

For information about modelling for earthquake conditions see the bibliography.

9.6 Fire

The most likely cause of damage to a structure in its lifetime is due to fire, but, as with seismic conditions, the modelling of the behaviour of a structure in a fire represents a major challenge in structural analysis. The input conditions in terms of fire load are difficult to predict and the stiffness and strength of structural materials – especially steel – significantly degrade as the temperature rises. Thermal movements of the structure, restraining actions and temperature gradients are important features of the behaviour of a structure in a fire.

Predictive modelling of the effects of a fire in a building is an emerging field and the 'model refinement principle' of Section 2.2.1 may need to be considered.

9.7 Temperature

9.7.1 General

While it is not common to include the effects of temperature in analysis models, temperature stresses can be high. It has been suggested, for example, that the highest stresses experienced by a building in its lifetime can be due to temperature stresses.

Typical situations where the effect of temperature may need to be included in the analysis model include:

- buildings where the expansion joints are further apart than normal or have not been provided
- bridges that do not have roller supports, i.e. integral bridges (O'Brien and Keogh 1999).

Provision for temperature changes and temperature gradients are often available in finite element software.

9.7.2 Basic relationships

The basic relationship for movement due to temperature change over a length L of a material is

$$\Delta = \alpha L T \qquad (9.1)$$

where Δ is the change in length over a distance L, α is the coefficient of thermal expansion of the material and T is the change in temperature.

9.8 Influence lines for moving loads

9.8.1 General

In bridge design in particular it is important to manipulate loads whose position on the structure is not fixed, i.e. moving loads. The influence line is a main technique used to identify critical positions of moving loads. Influence lines can also be used

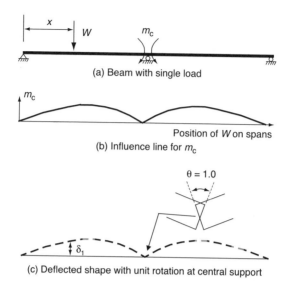

(a) Beam with single load

(b) Influence line for m_c

(c) Deflected shape with unit rotation at central support

Figure 9.2 Influence line for central support moment in a continuous beam.

to assess the best pattern of loading to use. Processing for influence lines is now done in software. This section discusses the fundamental principles behind influence lines.

9.8.2 Basic concept

An *influence line* is a plot where the abscissa is the position of a unit load on a structure and the ordinate represents a feature of behaviour that results from the application of the unit load – such as an internal action, a support reaction or a displacement. Consider, for example, the continuous beam in Fig. 9.2(a). If a single point load W traverses the structure left to right, the moment at the centre support, m_c, will:

- be zero when the load is at the left end
- increase to a maximum for a load at a position just past the centre of the first span
- reduce to zero when the load is over the central support.

The plot will be symmetrical about the centre support for a symmetrical system. This curve, shown on Fig. 9.2(b), is the influence line for support moment due to a transverse load on the beam.

9.8.3 Using influence lines

An influence line is used to find the highest value of an internal force action or displacement of a structure due to a load or set of loads moving across a structure. For example, in Fig. 9.2(a) it is clear that the position of the single-point load on the left-hand span that gives the greatest support moment is about 0.6 of the span from the left-hand end.

Normally, the loading to be applied is much more complex than a single-point load and so a process is needed to apply a set of loads to the influence line to calculate the maximum value of the variable. This is normally done in software.

9.8.4   Defining influence lines
The Mueller–Breslau method is used to calculate influence line curves as follows.

- *Internal force action* – a release is applied corresponding to the force action, and a unit displacement is applied at the release. The deflected shape of the structure which results from the unit displacement is the influence line for the internal action. This is illustrated in Fig. 9.2(c), where a unit rotation is applied corresponding to the central support moment. The resulting deflected shape is the same as for Fig. 9.2(b).
- *Reaction force* – the restraint corresponding to the reaction force is released and a unit deformation is applied to the structure in the direction of the reaction force. The shape of the deflected curve is the influence line for the reaction.
- *Displacement* – a unit point load is applied to the structure in the direction of the deformation, which is to be the ordinate of the influence line. The deflected shape of the structure under this load is the influence line for the displacement.

9.8.5   Validation information for the use of the Mueller–Breslau method for defining influence lines
The Mueller–Breslau method is based on the Maxwell–Betti reciprocal theorem, which is valid only for linear elastic systems. It cannot, therefore, be used with non-linear models.

9.9   Prestressing
Prestressing forces are applied along a line which is normally curved and eccentric to the centroidal axis of the member. Figure 9.3 shows a beam with end prestressing forces from a cable with profile as shown. The cable forces are specified as data and are converted to equivalent force actions at the nodes (O'Brien and Keogh 1999). The specification of the nodal loads is done by software. The user needs to define a number of factors, such as duct friction coefficient, relaxation loss factor, shrinkage coefficient, creep coefficient and end slip. Values for these parameters can be found in relevant codes of practice.

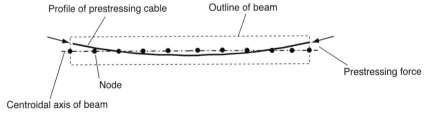

Figure 9.3   Prestressing force on a beam.

9.10   Impact loading

The modelling of impact loading tends to require high levels of non-linearity and is therefore mainly beyond the scope of this book.

9.10.1   Gravity impact

The forces involved in arresting the motion of a body falling under gravity depends on the stiffness of the arresting medium.

Elastic strain energy model

A simple model is to equate the kinetic energy of the body to the elastic strain energy at the point of impact. For example, if a mass M falls from a height h onto a simply supported beam (Fig. 9.4), the kinetic energy at impact will be equal to the potential energy at the beginning of the fall (neglecting frictional losses during the fall).

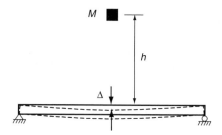

Figure 9.4   Mass falling on a beam.

Equate the kinetic energy to the potential energy and ignore the deflection of the beam in defining the latter results in the relationship

$$W = \sqrt{(2MghK)} \tag{9.2}$$

where W is the impact force on the beam, g is acceleration due to gravity, and K is the stiffness of the beam corresponding to a central point load, i.e. $W = K\Delta$ where Δ is the beam deflection. For equation (9.2) it is assumed that Δ is small compared with h.

Validation information

Equation (9.2) provides a simple means of estimating the effect of an impact loading. However, measurements to validate this approach tend not to give good correlation. This is likely to be due to non-linearities in the real impact. The model may only give indicative rather than predictive results.

10 Non-linear geometry

10.1 Introduction

This chapter discusses modelling methods for geometric non-linearity in general and methods of estimating elastic critical loads of skeletal frames and buildings.

10.1.1 Basic behaviour

As a structure is loaded, the geometry changes. *First-order theory* assumes that the effect of this change is unimportant in relation to prediction of displacement and internal forces in the system. *Second-order theory* takes into account the change in geometry and hence in the behaviour. The change in geometry causes the relationship between loads and displacements to be non-linear, and hence it is described as *geometric non-linearity*. (Note that the term 'second-order effect' is not exclusive to geometric non-linearity but can also refer, for example, to moments in truss members which are of secondary importance.)

As the load continues to increase, a situation may be reached where the internal restoring actions resulting from a destabilising displacement are just in equilibrium with the overturning effect of the applied load in relation to the displacement. This is the *critical load* condition – see Section 10.3.

Three basic types of geometric non-linearity are:

- *Large strains* – where large strains result in displacements that are not small in comparison with the dimensions of the system being analysed. This situation is very uncommon in structural engineering.
- *Small strains with finite displacements* – where the strains are small but the displacement of the structure is large enough to significantly affect behaviour. This occurs, for example, in the analysis of axially loaded slender structures and analysis of cable structures.
- *Critical load analysis* – where critical loads and corresponding mode shapes are treated as an eigenvalue problem (Section 10.3).

10.1.2 Cantilever strut example – the P-Δ effect

To demonstrate a non-linear geometry situation, Fig. 10.1 shows a cantilever strut supporting an axial load P. The system has a lateral deflection Δ (due to lack of straightness, lateral load or other source). This causes a moment, $P\Delta$, which will further increase the lateral deflection. This is known as the P-Δ effect – a type of geometric non-linearity that commonly needs to be taken into account in structural analysis. (In this book the symbol N is used for the vertical load on a structure

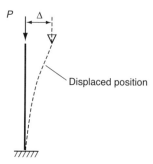

Figure 10.1 Axially loaded column.

– P is only used in Fig. 10.1 because the expression 'P-Δ effect' is well known.) In first-order theory, the effect of the $P\Delta$ moment is ignored.

Basic features of this type are:

- The effect of the eccentricity of the axial load in the members due to the bent shape of the member causes the displacements and internal actions to be magnified (Section 10.2).
- As the axial load P increases, the effective stiffness of the system decreases.
- Critical load N_{cr} (Section 10.3) may be reached, at which point the member can take no further axial load and the system is in a state of neutral equilibrium. The system may fail due to overstressing before the critical load is reached.

10.2   Modelling for geometric non-linearity

10.2.1   Using the non-linear geometry option in finite element packages

Most finite element packages provide a non-linear geometry option, by means of which second-order geometry effects are taken into account. It is normally quite easy to set this up, but the process of solution is iterative and convergence problems may arise. Most packages provide a range of solution options (e.g. total Lagrangian, Eulerian, etc.) and information about limits on the iteration process need to be specified. If convergence problems are experienced then experimentation with such features will be needed.

10.2.2   Use of the critical load ratio magnification factor

The effect of geometric non-linearity due to eccentricity of the axial load in members of a skeletal frame can be estimated using a magnification factor (Timoshenko and Gere 1961)

$$\eta = 1/(1 - \lambda) \tag{10.1}$$

where $\lambda = N/N_{cr}$ is the critical load ratio for the system, N is the axial load and N_{cr} is the elastic critical load.

Equation (10.1) implies that if (for example, for a frame analysed with non-linear geometry effects neglected) the moment in a particular member is M then the moment in the member taking account of the second-order effects will be ηM. For example, if $\lambda = 0.1$ then $\eta = 1/(1 - 0.1) = 1.11$, i.e. the moment (and the deflections) will be magnified by just over 11% (see example in Section 10.2.3).

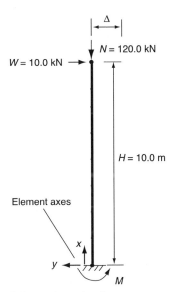

Figure 10.2 Cantilever with lateral top load.

Table 10.1 Properties for the cantilever of Figure 10.2

A_x: m^2	I_z: m^4	E: kN/m^2
0.123	8.0×10^{-5}	209×10^6

Note: A_x is the cross-sectional area.

10.2.3 Case study – non-linear geometry analysis of a cantilever

Figure 10.2 shows a cantilever with a top lateral load of $W = 10.0\,\text{kN}$ and a top applied axial load of $N = 120.0\,\text{kN}$. The cantilever was modelled using LUSAS[1] 3D non-linear thick beam elements (neglecting shear deformation) with eight elements in the height. The non-linear geometry function in LUSAS was invoked using the total Lagrangian solution option.

Restraints were applied in the z direction so that buckling would be in the x–y plane. The properties of the element are given in Table 10.1.

The critical load N_{cr} is (equation (10.5))

$$N_{cr} = \pi^2 EI_z/(2h)^2 = 3.142^2 \times ((209 \times 10^6) \times (8 \times 10^{-5}))/(2 \times 10)^2$$
$$= 412.7\,\text{kN}$$

where E is Young's modulus and I_z is the second moment of area of the section.

$$\lambda = N/N_{cr} = 120/412.7 = 0.291$$
$$\eta = 1/(1 - \lambda) = 1/(1 - 0.291) = 1.410$$

The first-order moment at the base of cantilever is

$$M_1 = 10 \times 10 = 100.0\,\text{kN}\,\text{m}$$

---

[1] LUSAS finite element modeller, FEA Ltd.

Table 10.2 Predictions of Δ_2 and M_2

1 Parameter	2 Using η	3 Non-linear element model	4 Difference: %
Δ_2: m	0.2811	0.2799	0.43
M_2: kN m	141.0	133.3	5.78

Notes: Difference (%) = (col. 2 − col. 3)/col. 3 × 100.

The second-order base moment using the magnification factor is

$$M_2 = M_1\eta = 100.0 \times 1.410 = 141.0\,\text{kN m}$$

The first-order top deflection in the line of the lateral load (Table A4) is

$$\Delta_1 = WH^3/(3EI) = 10 \times 10^3/(3 \times (209 \times 10^6) \times (8.0 \times 10^{-5})) = 0.1994\,\text{m}$$

The second-order top deflection using magnification factor is

$$\Delta_2 = \Delta_1\eta = 0.1994 \times 1.41 = 0.2811\,\text{m}$$

Table 10.2 compares the predictions of Δ_2 and M_2 from the magnification factor and from a non-linear geometry element model. This is a situation where the non-linear geometry effects are significant. The use of the magnification factor on the first-order values gives estimates that are higher but close to the values from the non-linear element model.

It is probable that the context of this example is especially favourable to the accuracy of the magnification factor approach; it is a simple structure, all parts of which are affected by the non-linear geometry behaviour. Accuracy of the magnification factor approach will not always be as good as in this case. For example, with local buckling the effect of the non-linearity will not be evenly distributed in the frame and the magnification factor may only apply locally.

10.2.4   Validation information for non-linear geometry effects

- *Criterion for neglecting non-linear geometry effects* – the criterion of expression (10.8) is often used to assess whether or not non-linear geometry effects need to be considered.
- *Design to code of practice* – code of practice rules for sizing of members under compressive axial load take account of second-order geometry effects. Therefore, if the structure is designed to the code of practice then an acceptance criterion for neglecting second-order effects may be "Design to code of practice number *xxx*". This may not provide a satisfactory validation if, for example, there are members in the structure that are more slender than normal.
- *Potential need to include non-linear material behaviour* – critical load analysis depends on elastic behaviour. Results from a model that incorporates non-linear geometry effects are potentially useful for understanding the behaviour of the structure. They may, however, not be predictive because elastic limits may be exceeded before a critical condition is reached, and non-linear material behaviour will also need to be taken into account.

- *Role of structural stiffness in non-linear geometry analysis* – apart from the level of the applied load, the main factor affecting the non-linear geometry effect is the stiffness of the system. This depends on several issues, including the E value, the connections, the restraints and the definition of the cross-sectional properties. Underestimating the stiffness will tend to be conservative in relation to non-linear geometry analysis.

10.3   Critical load analysis of skeletal frames

10.3.1   The Euler critical load for single members

Figure 10.3(a) shows a model of a uniform axially loaded strut with pin connections at both ends and an applied axial load N. Figure 10.3(b) shows a free body diagram of part of the strut. Taking moment equilibrium about the point a on Fig. 10.3(b) gives

$$Nv + M = 0.0 \tag{10.2}$$

where N is the axial force in the strut, v is the lateral displacement at point a and M is the internal moment at point a.

The positive direction of M is shown in Fig. 10.3(b). It has a negative value for equation (10.2).

Substituting for $M = EI \mathrm{d}^2 v/\mathrm{d}x^2$ (equation (5.1)) gives

$$Nv + EI \mathrm{d}^2 v/\mathrm{d}x^2 = 0.0 \tag{10.3}$$

Equation (10.2) exemplifies the critical load situation. Nv is an 'overturning' moment due to the eccentricity of the applied load tending to bend the strut and $M = EI \mathrm{d}^2 v/\mathrm{d}x^2$ is a 'restoring' moment provided by the elastic strain energy in the material tending to straighten the strut.

When Nv and M have the same value, the system is in a state of unstable equilibrium – the critical load state.

Solution of equation (10.3) for the strut with pinned ends gives the elastic critical load for this condition

$$N_{\mathrm{cr}} = \pi^2 EI/L^2 \tag{10.4}$$

where L is the length of the strut and EI is the bending stiffness parameter.

Equation (10.4) gives the *Euler critical load* for a pin-ended strut.

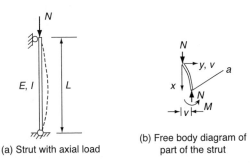

(a) Strut with axial load

(b) Free body diagram of part of the strut

Figure 10.3   Pin ended strut.

Table 10.3 Effective length factors for struts

End conditions	c_e
Restrained in position but not restrained in direction at both ends (pin–pin)	1.0
Fixed both ends (fully fixed)	0.7
Fixed one end, free at other end (cantilever)	2.0
Fixed in position but partially restrained in direction at both ends	0.85

For struts with other end conditions, the critical load can be calculated using

$$N_{cr} = \pi^2 EI/(L_e)^2 \tag{10.5}$$

where $L_e = c_e L$ is the *effective length* of the strut and c_e is an effective-length factor which depends on the end conditions. Typical values for c_e are given in Table 10.3.

10.3.2   Non-sway instability of a column in a frame

Non-sway instability occurs when a column buckles without any lateral movement at its ends. Figure 10.4 shows a frame with a lower storey column which is significantly more flexible in bending than the other columns of the frame. The buckled shape from an eigenvalue extraction run on the frame shows how this column buckles in a local mode. This type of behaviour is well covered in codes of practice, but a quick check on the potential for non-sway instability of a member can be made using expression (10.8) and estimating N_{cr} using equation (10.5).

10.3.3   The critical load ratio for an axially loaded member of a frame

The critical load ratio is defined either as

$$\lambda = N/N_{cr} \tag{10.6}$$

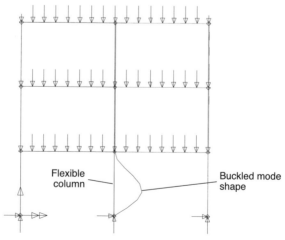

Figure 10.4 Non-sway buckling of a column.

or as

$$\lambda_r = N_{cr}/N \tag{10.7}$$

where N is the axial load in a member and N_{cr} is the elastic critical load for the member.

Unfortunately, there is no standard for defining the critical load ratio. Both λ and λ_r (the r subscript denotes 'reciprocal' and is introduced here to differentiate between the two definitions) are used in codes of practice.

λ is a main parameter for assessing the potential effect of non-linear geometry – $\lambda = 1.0$ corresponds to the critical load state. It is best to keep well away from this situation at working load, and the criterion

$$\lambda < 0.1 \tag{10.8}$$

is commonly used in codes of practice for this purpose.

10.3.4   Estimation of critical loads using eigenvalue extraction

A structure can buckle in a range of modes. In critical load analysis of frames, the number of potential critical loads is equal to the number of degrees of freedom in the element model. (This is an analogous situation to natural frequency analysis in dynamics (see Section 11.3.1) – the number of potential natural frequencies is equal to the number of degrees of freedom.) This is an eigenvalue situation. To each value of critical load (the eigenvalue) there corresponds a buckling mode shape (the eigenvector).

The outputs from such a process are a list of critical load ratios and the buckling mode shapes. The software normally allows the user to define the number of eigenvalues to be defined. It is important to ensure that if a critical load is used for design then it corresponds to the relevant mode shape.

10.3.5   Case study – eigenvalue analysis of a cantilever strut

An eigenvalue buckling analysis was carried out for the cantilever column shown in Fig. 10.5 using LUSAS 13.3[2]. The axial load applied in the loadcase was 100.0 kN. Figure 10.5 shows the first two in-plane buckling modes. The critical load for the first mode is that given by the Euler formula, which, using equation (10.5), is

$$N_{cr} = \pi^2 EI/(2L)^2$$

Using $N = 100.0$ kN, $E = 209 \times 10^6$ kN/m^2, $I_z = 0.0004025$ m^4 and $L = 4.0$ m (as in the eigenvalue analysis) gives

$$\lambda_r = N_{cr}/N = (\pi^2 EI/(2L)^2)/N$$

$$= (3.14159^2 \times (209 \times 10^6) \times (0.0004025)/(2 \times 4)^2)/100 = 129.727$$

The number of elements in the model was varied, giving the results shown in Table 10.4. Note that:

---

[2] LUSAS finite element modeller, FEA Ltd.

- the accuracy of the eigenvalue extraction, for the type of solution used in the software, is a function of the number of elements used within the height of the strut
- the critical load ratio converges to the Euler value.

The predicted critical load ratio for the second in-plane mode is 1170.3, which represents a natural frequency that is about nine times that of the first mode. There are also out-of-plane modes, which are not considered here.

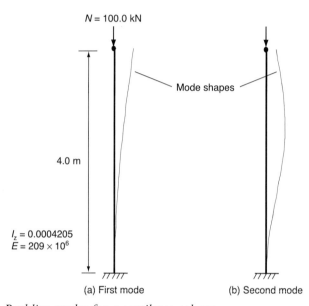

Figure 10.5 Buckling modes for a cantilever column.

Table 10.4 Elastic critical load ratios from eigenvalue analysis

Number of elements	λ_r	Difference from Euler value ($= 129.727$): %
40	129.761	0.026
10	130.266	0.415
5	131.892	1.669
4	133.136	2.628
3	135.894	4.754
2	144.331	11.257

10.4   Global critical load analysis of building structures

The traditional approach to stability analysis of buildings is to consider critical load on a column-by-column and a frame-by-frame basis.

Codes of practice require a decision to be made as to whether or not a frame is a 'sway' or a 'non-sway' frame. A non-sway frame is one for which non-linear geometry effects can be neglected. Frames are assessed individually for this condition, but if a lateral instability event were to occur for a building it could not affect only a single frame. The whole building structure (or at least a major part of it) must be involved. Taking the whole building structure into account in the assessment of sway and torsional critical loads is the basis of the global critical load approach (MacLeod and Zalka 1996, Zalka 2002).

The *global critical load* is the lowest critical load for the building as a whole (including torsion). An estimate of the global critical load of the building is made and if this is sufficiently greater than the total vertical load on the building then non-linear geometry effects may be neglected for the building as a whole, although local instabilities may still need to be considered.

The eigenvalue extraction process for critical load analysis described in Section 10.3.4 is now a very realistic option which will tend to give the most accurate predictions. The equivalent column approach, for which the most comprehensive version is to be found in Zalka (2002), can be used as a checking model and for design in appropriate situations.

11 Dynamic behaviour

11.1 Introduction

Two basic issues in dynamic analysis are:

- the calculation of natural frequencies of structures to deal with a range of situations, including resonant frequency analysis – this chapter deals with this issue.
- the prediction of response of structures to inputs which are not cyclic – this involves *time stepping* (where the dynamic input is broken down into small time steps), which is considered to be advanced analysis and so is not addressed here.

For more detailed information on modelling for dynamic effects, see the bibliography.

11.2 Dynamic behaviour of a single mass and spring system

11.2.1 Governing equation

Figure 11.1 shows a model of a mass that is free to move horizontally and is restrained by a spring and a viscous damper. The equation of motion of this system is

Inertia   Spring
force   force

$$M\frac{d^2u}{dt^2} + C\frac{du}{dt} + Ku = P(t) \qquad (11.1)$$

Damping   Forcing
force   function

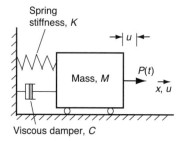

Figure 11.1 Single degree of freedom system.

Equation 11.1 is an equilibrium equation and has the following characteristics.

- The behaviour is defined by a single degree of freedom in the x direction, where the deformation in the x direction is u.
- The mass M (kg) develops an inertia force that is proportional to its acceleration.
- The damping force is proportional to the velocity of motion – C (kN/(m/s) = kg/s) is the damping constant.
- The linear spring has stiffness K (kN/m = kg/s^2)
- A force $P(t)$ (kN), which is a function of time, is applied to the system.

11.2.2   Validation information for equation (11.1)
Validation issues in relation to the terms of equation (11.1) include the following:

- The *inertia force*, $M \, \mathrm{d}^2u/\mathrm{d}x^2$ – this depends on Newton's second law of motion, which is accurate at conventional velocities. The main validation issue is the degree of accuracy in defining the mass M. The dead-load mass of the structure can normally be estimated to reasonable accuracy, but the mass due to live load may be much more approximate.
- The *damping force*, $C \, \mathrm{d}u/\mathrm{d}t$ – to assume that the damping force is proportional to velocity is to assume *viscous damping* and that the damper itself is a *dashpot*. For dampers such as shock absorbers on cars, viscous damping may be quite accurate, but for structural systems the damping is more a function of friction than of velocity. Therefore, the viscous model is assumed more for convenience than for accuracy. It does not represent the real behaviour well, but nevertheless gives results that are useful for assessing the degree of damping of a structural system.
- The *spring force*, Ku – the validity of this depends on the several assumptions made in establishing the stiffness for the analysis model, such as those in relation to the (dynamic) E value, supports, connections, cracking in concrete, etc.
- The *forcing function*, $P(t)$ – the accuracy here can vary from being good in relation to vibration due to machinery to very approximate for wind, earthquake and blast loading.

An important issue is that a square root (or the equivalent of a square root for multi-degree of freedom systems) is used to find the natural frequency – see equation (11.2). This has positive and negative features.

- Taking the square root decreases the sensitivity of the natural frequency to changes in mass and stiffness, thus helping to improve the accuracy of the predictions.
- Due to this lack of sensitivity, it can be difficult to make design changes to alter significantly the natural frequency.

For natural frequency calculations in structural engineering, damping may not to be important (Section 11.2.4) and therefore the main issues for free vibration tend to be mass and stiffness. It may be possible to calculate the mass with reasonable accuracy, but the real stiffness of a structure can be difficult to model accurately.

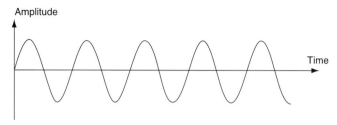

Figure 11.2 Free vibration – no damping.

The effects of features such as partial connection fixity, *EI* values for concrete and the contribution of non-structural components can have a high degree of uncertainty.

11.2.3 Free undamped vibration

Figure 11.2 shows the situation when $C = 0.0$ and $P(t) = 0.0$, i.e. *free undamped vibration*. This would occur if the system with no damping was pulled to one side and released, i.e. the equivalent of plucking the string of a guitar. The system vibrates at a natural frequency which is independent of the amplitude.

Solution of equation (11.1) for this situation gives:

- The undamped natural circular frequency ω_n

$$\omega_n = \sqrt{\frac{K}{M}} \text{ (radians/s)} \tag{11.2}$$

- The undamped natural frequency f

$$f = \omega_n/2\pi \text{ (cycles per second)} \tag{11.3}$$

11.2.4 Damping

There is always some damping, which causes the amplitude to decay to zero, and therefore the plots of amplitude against time in Fig. 11.3 are more realistic. The natural circular frequency with damping ω_1 is given by

$$\omega_1^2 = \frac{K}{M} - \left(\frac{C}{2M}\right)^2 \tag{11.4}$$

The following situations are identified.

1. $C = 0.0$ – this is the undamped situation (Fig. 11.2).
2. $K/M > (C/2M)^2$ – this is the damped free vibration situation. The response of the system is as shown in Fig. 11.3(a). The system oscillates at a natural frequency but this motion decays exponentially. This is the normal situation in structural systems.
3. $K/M < (C/2M)^2$ – this is the *overdamped* situation. The damping is so high that when the system is 'plucked' it does not go back beyond the static position (Fig. 11.3(b)).
4. $K/M = (C/2M)^2$ – this is the *critical damping* situation, which occurs at the boundary between situations 2 and 3.

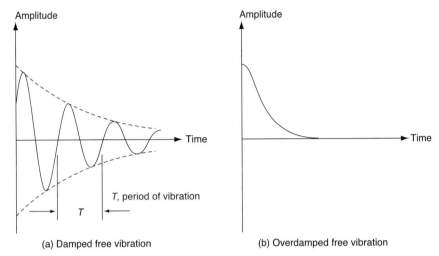

(a) Damped free vibration   (b) Overdamped free vibration

Figure 11.3 Damped and overdamped situations.

The damping constant at critical damping C_c is given by

$$C_c^2 = 4MK \tag{11.5}$$

It is common to express the degree of damping in relation to the critical value using the *damping ratio*

$$\zeta = C/C_c \tag{11.6}$$

Substituting $C = \zeta C_c$ into equation (11.4) and using equation (11.5) gives

$$\omega_1 = \omega_n \sqrt{1 - \zeta^2} \tag{11.7}$$

Equation (11.7) gives the natural circular frequency for systems under viscous damping. In structural systems, the damping ratio ζ is typically 5%. Substituting this into equation (11.7) gives $\omega_1 = 0.9987\omega_n$, thus levels of damping of this order have a negligible effect on the natural frequency. Therefore, for natural frequency calculations for structural systems, the effect of damping is normally neglected and equation (11.2) is used rather than equation (11.4).

11.3   Multi-degree of freedom systems

11.3.1   Basic behaviour

Consider the plane frame cantilever model in Fig. 11.4(a), which has a mass of 100 kg at the top and at mid-height The mass of the cantilever itself is neglected in this context. A dynamic analysis of this system gave the first two mode shapes shown in Figs 11.4(b) and (c).

The natural frequencies are 15 and 99 Hz respectively. When vibrating at 15 Hz, the shape of the deformation is as shown in Fig. 11.4(b). This is the *mode shape*, which is independent of amplitude (the natural frequency is independent of amplitude). When the cantilever vibrates at 99 Hz it does so with the mode shape

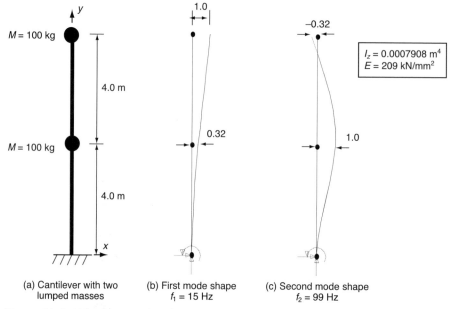

Figure 11.4 Vibration modes for a cantilever.

shown in Fig. 11.4(c). There is a unique mode shape of vibration for each natural frequency.

The free vibration situation represents an *eigenvalue* problem. The mode shapes are the *eigenvectors*. For the dynamic problem an eigenvalue is ω_i^2, which is converted to the natural frequency f_i using equation (11.3). The solution process for the eigenvalue problem is known as *eigenvalue extraction*.

In a multi-degree of freedom system, such as that in Fig. 11.4, the number of potential eigenvalues is equal to the number of degrees of freedom for the model. The eigenvalue extraction process normally identifies eigenvalues in ascending order, starting with the lowest (which tends to be the most important).

For the model in Fig. 11.4 there are two plane frame nodes – at the top and at mid-height. There are three degrees of freedom at each node (two translations and a rotation) and therefore there can be six eigenvalues (and associated eigenvectors). There are two lateral freedoms, giving the natural frequencies (derived from the eigenvalues using equation (11.3)) and the mode shapes (the eigenvectors). There will also be higher eigenvalues, which will be relevant to the vertical freedoms and the rotational freedoms.

11.3.2 Governing equation for multi-degree of freedom systems

The following relationship is a matrix notation equivalent of equation (11.1) for multi-degree of freedom systems

$$[M]\{\ddot{u}\} + [C]\{\dot{u}\} + [K]\{u\} = \{P(t)\} \tag{11.8}$$

where M is the mass matrix, C is the damping matrix, K is the stiffness matrix, $\ddot{u}$ is the acceleration vector, $\dot{u}$ is the velocity vector, u is the displacement vector and $P(t)$ is the forcing function vector.

The validation information for equation (11.1) given in Section 11.2.2 is also relevant to equation (11.8).

For undamped free vibration, equation (11.8) reduces to

$$[M]\{\ddot{u}\} + [K]\{u\} = 0.0 \tag{11.9}$$

11.3.3   Modelling for dynamic eigenvalue extraction

Using modern structural analysis software, the extraction of eigenvalues is straightforward. The model is set up and the dynamic eigenvalue option is selected. The number of required eigenvalues is set from the data and the system will output the natural frequencies and mode shapes. In some cases the extraction process may be unstable. To overcome this, the system may provide options for the type of eigenvalue solver. Experimentation with these may be needed.

When dealing with the results of eigenvalue extraction it is important to associate the natural frequency with the mode shape. For example, in the 3D analysis of a building it may be necessary to investigate the natural frequency of the floor slabs. The mode shapes need to be viewed one at a time until the floor vibration modes are identified. The frequency that corresponds to these modes is then the one to consider in relation to floor vibration.

11.3.4   Verification of output for dynamic models

Care is needed to ensure that the mass has been correctly input. A check that the mass used by the system is that intended to be used is worthwhile. If special elements have been added to simulate high but finite stiffness to create rigid links (Section 4.5.2), it is important to make sure that these elements do not add non-relevant mass to the system.

A visual check on the mode shapes (eigenvectors) to make sure that they correspond with expectations is essential.

Checking models for natural frequencies are discussed in Section 11.6.

11.4   Resonance

11.4.1   Description

When the frequency of the forcing function and a natural frequency of the system are close, then they reinforce each other and the system goes into *resonance*. This can happen not only with the lowest natural frequency but also with higher natural frequencies.

For example, with suspended floors used for dancing, if the rhythm of the dancers is at or close to a natural frequency of the floor (normally the lowest mode of vibration in this case) then the deflections of the floor will be amplified causing discomfort to the dancers and possibly damage to the structure.

11.4.2   Systems subject to vibratory loading

If the structure is subject to an applied oscillating loading, such as for machinery supports, the applied load can be defined by

$$P(t) = P_0 \sin \Omega t \qquad\qquad (11.10)$$

where Ω is the 'forcing frequency' and P_0 is a force.

The frequency ratio Ω/ω_n is useful for characterising the behaviour of the system (ω_n is the natural circular frequency).

Figure 11.5 shows the relationship between the frequency ratio and the dynamic load factor (DLF), which is the ratio of the maximum displacement of the system under dynamic conditions to the displacement of the system with the load P_0 applied statically. This shows that:

- with low frequency ratio ($\ll 1.0$) the system has time to respond fully to the applied load and the deflections and the stresses are as if the load is static
- with high frequency ratio ($\gg 1.0$) the system does not have time to respond to the very fast changes in load and the forces in the system can be low (less than P_0)
- at, or close to, resonance ($\Omega/\omega_n = 1.0$) the forces in the system are magnified, sometimes heavily – this effect is mitigated by damping.

When trying to ensure that the lowest natural frequency of the system is greater than a given value (e.g. greater than 4 Hz for a floor system), it will be conservative to underestimate the stiffness and overestimate the mass. When trying to identify resonance for natural frequencies greater than the lowest, it will not be practical to define a conservative direction in which to make assumptions. The most accurate model has to be used.

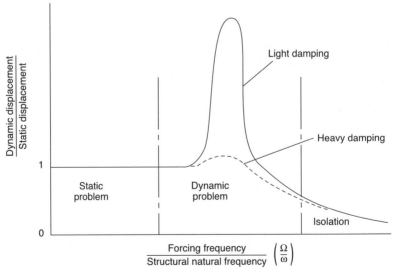

Figure 11.5   Effect of ratio of forcing to natural frequencies.

11.5 Transient load

It is possible by using appropriate software to carry out a 'time-stepping' analysis, where a dynamic load is applied in short time intervals and the dynamic response is modelled. In the context of a pulse of load being applied to a structure, an important factor is the ratio of the natural period of the system ($T_s = 1/f$) and the time of the pulse T_p. If T_p/T_s is significantly greater than 1.0 then it may be possible to treat the pulse as a static load. If T_p/T_s is significantly less than 1.0 then the effect of the pulse may be negligible.

11.6 Checking models for natural frequencies

11.6.1 Single-span beams

The natural frequency f_1 of single-span beams can be estimated using equation (11.11)

$$f_1 = \frac{a}{2\pi} \sqrt{\frac{EIg}{WL^3 + bwL^4}} \tag{11.11}$$

where a and b are coefficients (Table 11.1), W is a central point load or end load for a cantilever (kN) (mass of system $= W/g$), w is the uniformly distributed load over the length of beam (kN/m) (mass of system wL/g), L is the span, I is the second moment of area and g is acceleration due gravity.

11.6.2 The maximum deflection formula

The following approximate formula for the natural frequency of a beam system is given in Wyatt (1988)

$$f = 18/\sqrt{\delta} \tag{11.12}$$

where f is the natural frequency (Hz), δ for a single beam is the maximum deflection (mm) of the beam loaded by the mass that it supports (converted to weight, i.e. the applied load is $wL = Mg$ where M is the mass of the beam) and δ for a multi component structural system (e.g. with frames, slabs, etc.) is the maximum deflection (mm) of the system under an applied load of magnitude equal to the weight that corresponds to the mass of system and in direction to that of the shape of the mode of vibration being considered (see the case study in Section 11.6.3).

Equation (11.12) is based on equation (11.11), with $W = 0.0$ and the corresponding beam deflection formula from Table A4. This expression can give good estimates of natural frequency for systems with distributed loading.

Table 11.1 Beam coefficients

Case	a	b
Simply supported	6.93	0.486
Fully fixed	13.86	0.383
Cantilever	1.732	0.236

11.6.3   Case study – use of equation (11.12)

Figure 11.6 shows a beam element model of a two-span continuous beam. The beam is a $254 \times 146 \times 31$ kg/m universal beam (UB). Considering the vibration of the beam only (no added mass), the applied load to be used is

$$mg = 31 \times 9.8 = 304\,\text{N/m}$$

where m is the mass per metre.

This load is applied vertically downwards in the left-hand span and vertically upwards in the right-hand span to correspond to the first mode of vibration (shown in Fig. 11.6).

From the run using this as static loading, the maximum deflection in the left-hand span is 0.1732 mm. Using this in equation (11.12) gives

$$f = 18/\sqrt{0.1732} = 43.25\,\text{Hz}$$

Using eigenvalue extraction, the first mode frequency is

$$f = 43.84\,\text{Hz}$$

The difference between the two values for f is 1.3%.

11.6.4   Single mass and spring

If the system can be modelled as a single mass and spring then equation (11.2) can be used. For example, Fig. 11.7(a) shows dead load W centrally placed on a simply supported beam. Figure 11.7(b) shows the equivalent mass and spring.

The mass is $M = W/g$ and the spring stiffness is the load to cause unit deflection in the line of W. This can be obtained by manipulating the relevant deflection coefficient from Table A4, i.e. from this table

$$\Delta = WL^3/(48EI)$$

hence

$$K = W/\Delta = 48EI/L^3$$

Therefore, using equation (11.3) gives

$$f = 1/(2\pi)\sqrt{(K/M)} = (1/(2\pi))\sqrt{((48EI/L^3)/(W/g))} = 1.102\sqrt{(EIg/WL^3)}$$

This is the same value as using equation (11.8).

With distributed mass, approximations will be needed, e.g. with uniformly distributed mass 50% of the mass might be assumed to act as a central mass.

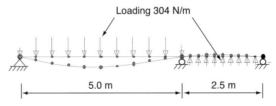

Loading 304 N/m

|← 5.0 m →|← 2.5 m →|

Figure 11.6 Continuous beam.

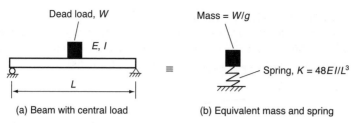

Figure 11.7 Single spring model for beam with central load.

11.6.5 Combinations of frequencies

Component frequencies due to different load cases can be approximately combined using the Dunkerly method

$$\frac{1}{f_0^2} = \frac{1}{f_1^2} + \frac{1}{f_2^2} + \cdots \tag{11.13}$$

where f_0 is the combined or system frequency and f_1, f_2 etc., are the natural frequencies of the components. The accuracy of this process depends on the mode shapes for the separate frequencies being similar.

12 Case studies

This chapter provides two examples which show how structural analysis may be carried out in the modern context.

12.1 Case study 1 – vierendeel frame

12.1.1 General

This example is for a simple plane frame and demonstrates a range of modelling issues and activities. For a structure of this type, a detailed modelling review as set out here may not be needed if it is commonly used. However, the results verification should be carried out as a standard procedure.

The examples should be read in conjunction with the description of the modelling process given in Chapter 3.

12.1.2 Definition of the system to be modelled – the engineering model

Portrayal

Figure 12.1(a) shows a plan of a building structure which incorporates a vierendeel roof truss arrangement, fabricated using hot rolled steel sections. Figure 12.1(b) shows an elevation of a frame and Figs 12.1(c) and (d) show details of a column baseplate support and the beam-to-column connection, respectively.

Requirements of the model

The requirements are to estimate the displacements and internal force actions due to dead load and vertical live roof load on a frame.

12.1.3 Model development

2D or 3D model?

A 3D model of the system would give better accuracy than a 2D model of a single frame, but the latter is normally used and is adopted here to simplify the example for demonstration purposes.

Elements and mesh

While it would be possible to model the frame using flat shell elements, the only realistic option is to use beam elements. Using engineering beam elements (Section 5.5.2) there are no discretisation issues in defining the model, i.e. there is no need for mesh refinement.

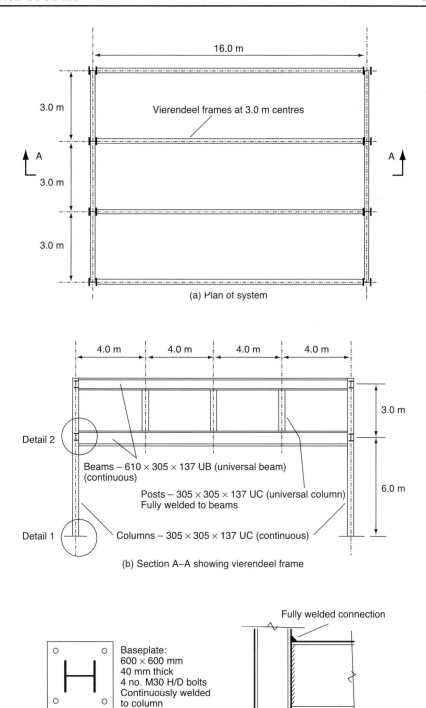

Figure 12.1 Vierendeel frame structure.

Material model
The options are to use either an elastic material model or a model that allows plastic hinges to form. This context could be realistic for the latter type of model, but the elastic model is chosen because of the availability of software.

Supports
The detail at the base of the column can take a moment, but pinned supports are used in the first instance. For the vertical loading, the sensitivity analysis (Section 12.1.7) shows that the difference between pinned and fixed supports is negligible in this case. This matter is addressed in the validation analysis (see Table 12.2).

Connections
The connections are all capable of being designed as having full moment continuity so they are modelled as such. Even with the full welding there may be some local rotation due to flange distortion, but full moment continuity is a conventional assumption for a frame of this type.

The finite sizes of the member could be included in the model, as discussed in Section 5.6.2, but neglecting this will be conservative for displacements.

Loading
The loading is not specified for this context, but the uncertainty in the dead loading will be low and the code values for live loading are likely to be conservative.

12.1.4 The analysis model
Figure 12.2 shows the analysis model.

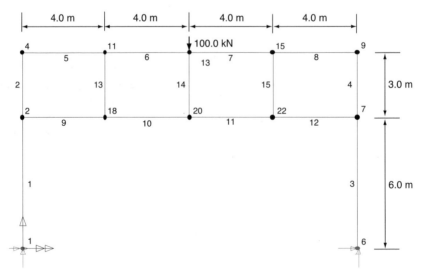

Figure 12.2 Vierendeel frame analysis model.

Table 12.1 Element properties for vierendeel frame analysis

Part	A_x: m^2	A_y: m^2	I_z: m^4	E: kN/mm^2
Column, posts	0.0174	4.42	0.0003281	209
Beams	0.019	7.23	0.001259	209

Analysis program
The software selected for the analysis was LUSAS 13.3.[1]

Units
Units used are metres and kilonewtons.

Elements
The elements are thick beam in-plane elements with shear deformation neglected. Properties are given in Table 12.1.

Support conditions
Nodes 1 and 6 are pinned.

Connections
All connections are assumed to have full moment continuity. The effect of finite widths of members at the connections (Fig. 5.23) is neglected.

Non-linear geometry
Non-linear geometry effects are neglected.

Loading
A nominal checking load of 100.0 kN is applied vertically downwards at node 13.

12.1.5   Model validation
The validation analysis is set out in Table 12.2.

12.1.6   Results verification
Error warnings from software
There were no error warnings.

Data check
No errors were identified.

Sum of support reactions
Table 12.3 gives the support reactions which can be accepted.

[1] LUSAS Finite Element Modeller, FEA Ltd.

Table 12.2 Validation analysis for vierendeel frame model

Modelling issue	Acceptance criterion	Outcome
Linear elasticity	Provides a set of internal actions which can be justified for ultimate load design on the basis of the lower bound theorem (Section 2.3.3). Design to code of practice.	LSR
Bending theory, shear deformation	Criterion: $L/d > 10$ L/d posts $= 3/0.305 = 9.8$ L/d beams $= 4/0.610 = 6.6$ On the basis of the sensitivity analysis (Section 12.1.7) include shear deformation (if software allows it).	Amend
Finite size of connections neglected	Conventional assumption; conservative for estimations of displacement. Use beam moments at faces of columns for design.	CA/LSR
Rigid moment connections	Design to code of practice	LSR
Pinned supports	Better to use a rotational stiffness of $4EI_c/h$ (BS 5950)	Amend
Non-linear geometry effect neglected	Design to code of practice	LSR

Notes: LSR – later stage requirement; CS – criterion satisfied; CA – conventional assumption.

Table 12.3 Support reactions for vierendeel frame

Node	F_x	F_y	M_z
1	0.2372266714529	5.0	0.0
6	−0.2372266714529	5.0	0.0
Sum	0.0000000000000	10.0	0.0

Restraints

Table 12.4 shows that the pinned supports at nodes 1 and 6 have been properly implemented.

Symmetry check

The values quoted in Table 12.3 show that the symmetry condition is satisfied (to 13 significant figures).

Table 12.4 Support deformations for vierendeel frame

Node	Δ_x	Δ_y	θ_z
1	0	0	2.25×10^{-5}
6	0	0	-2.25×10^{-5}

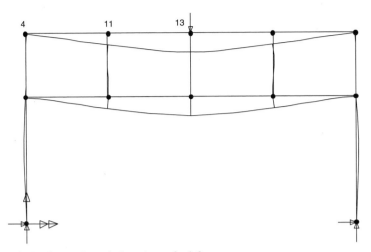

Figure 12.3 Deformed mesh for vierendeel frame.

Qualitative check – deformations
Figure 12.3 shows the deformed mesh for the checking load. The vierendeel frame will deform in a dominant shear mode (see Section 5.10.3 and the sensitivity analysis in Section 12.1.7) which for the checking loadcase will give a straight line deflection from the centre of the span to the column. The displaced position of nodes 4, 11 and 13 are close to being in a straight line. The bending deformation of the posts does not show in Fig. 12.3, but the columns bowing outwards are consistent with the rotation of the end of the lower beams. There are no negative observations.

Qualitative check – internal force actions
Figure 12.4 shows the bending moments in the frame taking the checking load. Where the bending moment line crosses the longitudinal axis of the element is a

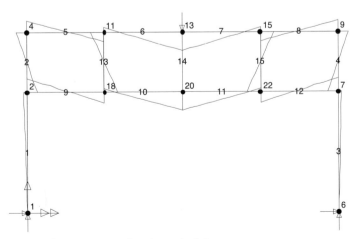

Figure 12.4 Bending moments for vierendeel frame.

point of contraflexure. Having points of contraflexure close to the mid-lengths of members is a characteristic of vierendeel frames. In the case of Fig. 12.4 this tends to be so, except for elements 6 and 10 (and the corresponding symmetrical pair 7 and 11). These elements have the point of contraflexure further away from the mid-positions because:

- the posts are relatively flexible in bending as compared with the beams, i.e. the ψ value is not low
- the ends of elements 6 and 10 at the centre of the span (i.e. at nodes 13 and 20) are, in effect, fully fixed, i.e. there is no joint rotation for this loadcase. This localised stiffness causes the moment to be drawn towards these connections, pushing the points of contraflexure away from the connections.

The moments in the columns are low because of their relatively low bending stiffness (I/L value) as compared with the beam.

There are no negative observations.

Checking model – deformations

The vierendeel frame is treated as an equivalent beam (Fig. 12.5) – as described in Section 5.11.3. For this model the beams are denoted as *chords*.

The mid-span deflection of the equivalent beam Δ_{frame} using equation (5.22) is

$$\Delta_{\text{frame}} = \Delta_b + \Delta_s$$

where Δ_b is the deflection due to the bending mode effect of the axial deformation of the chords and Δ_s is the deflection due to the shear mode effect of the bending of the chords and posts.

In what follows the subscript 'c' denotes *chord*. Using the expressions for Δ_b and Δ_s from Table A4 and equations (5.16) and (5.24)

$$\Delta_{\text{frame}} = WL^3/(48E_cI_g) + WL/(4K_{sv})$$

$$I_g = A_cb^2/2 = 0.019 \times 3^2/2 = 0.0855$$

$$\psi = (I_c/a)/(I_p/b) = (0.001259/4.0)/(0.0003281/3) = 2.88$$

$$K_{sv} = 24E_cI_c/(a^2(1+2\psi)) = 24 \times (209 \times 10^6) \times 0.001259/4.0^2(1+2 \times 2.88)$$

$$= 58387\,\text{kN}$$

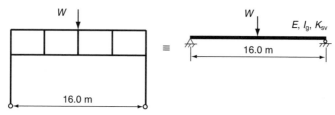

Figure 12.5 Equivalent beam model for vierendeel frame.

hence

$$\Delta_{\text{frame}} = 100.0 \times 16.0^3/(48 \times (209 \times 10^6) \times 0.0855)$$
$$+ 100.0 \times 16.0/(4 \times 58387)\text{m}$$
$$= 0.476 + 6.77\,\text{mm}$$
$$= 7.25\,\text{mm}$$

Element model value Δ_{em}

$$\Delta_{\text{em}} = 3.97\,\text{mm}$$

$$\text{percentage difference} = (\Delta_{\text{frame}} - \Delta_{\text{em}})/\Delta_{\text{em}} \times 100$$
$$= (3.97 - 7.25)/3.97 \times 100 = -82.6\%$$

Reasons for the difference between the two values include:

- The equivalent beam model for shear stiffness assumes points of contraflexure at the mid-lengths of all members (Section 5.11.3). This is equivalent to inserting pins into the structure to make it more flexible. The greater the real positions of the points of contraflexure deviate from the mid-length positions, the greater will be the overestimate of deflection by the equivalent beam. It appears that the points of contraflexure not being close to the centre of the mid-lengths of the chord panels in the centre of the span makes a big difference to the accuracy of the estimation of Δ_s. If the flexural stiffness of the posts is made significantly larger than that of the beams (i.e. ψ significantly less than 1.0) then the correlation between the checking model and the element model would be much better.
- The columns provide rotational and horizontal restraints at the supports, which will cause the real structure to be stiffer than the simply supported equivalent beam – see sensitivity analysis in Section 12.1.7 which shows that this is not an important issue.

The correlation between the element model and the checking model is not good, but the difference can be explained.

Checking model – internal force actions

The moments in the beam members at the centre of the span are estimated based on an assumption about the point of contraflexure in their lengths.

The element model results for the top and bottom beam elements at the centre of the span are given in Table 12.5.

Table 12.5 End actions for elements 6 and 10

Element	Node	N_x: kN	S_y: kN	M_z: kN m
6	11	73.3	26.2	19.8
6	13	−73.3	−26.2	85.1
10	18	−71.0	23.8	14.5
10	20	71.0	−23.8	80.7

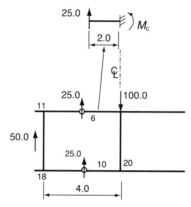

Figure 12.6 Free body diagram to calculate shear forces at centre span.

Figure 12.6 shows a free body diagram of the centre part of the truss. The shear across the frame is constant at 50.0 kN. Assuming that this is shared equally by the two chord members gives 25.0 kN in each chord. Assuming a point of contra-flexure at the mid-length of the panel between the posts gives the free body diagram for the chord in the panel as shown in Fig. 12.6. On this basis, the moment in the chord at the centre of the frame is

$$M_c = 25.0 \times 2.0 = 50.0 \text{ kN m}$$

The element model value for the top chord (element 6) from Table 12.5 is

$$M_c = 85.1 \text{ kN m}$$

Two factors cause the moment to be underestimated by the checking model.

- The points of contraflexure for elements 6 and 10 are not at the mid-length of the chord panel (as in Fig. 12.4) but at a position 3.24 m from the centre of the frame (in the case of the top chord).
- The shear in the top chord is greater than that in the bottom chord (due mainly to the horizontal restraining action of the columns).

Using the value of 26.2 kN for the shear in element 6 from Table 12.5 gives the moment in the chord at the centre of the frame as $26.2 \times 3.24 = 84.9$ kN m. The small difference between this value and the value of 85.1 in Table 12.5 is due to the low number of significant figures quoted in the table.

Again the correlation between the checking model and the element model results is not close, but it can be explained. The checking model used tends to be less accurate with a small number of panels.

Review of verification outcomes
There are no negative observations in the verification. The correlation between the element model and the checking model is not good, but the differences can be explained. Accept the model at this stage.

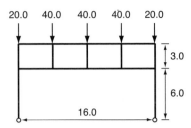

Figure 12.7 Loading for sensitivity analysis of vierendeel frame model.

12.1.7 Sensitivity analysis

For the sensitivity analysis, a more realistic roof loading of 10.0 kN/m is applied vertically on the top beam. Figure 12.7 shows the equivalent nodal loads used for this loading.

Feature variation

A *reference model* (Section 2.4.4) – model 1 in Table 12.6, based on the system shown in Fig. 12.4 was used with the following features:

- pin supported columns
- no shear deformation of elements
- loading as in Fig. 12.7.

The *indicative parameters* (Section 2.4.4) used are:

- Δ_{13} – the central vertical deflection of the frame at the top beam level
- S_6 – the shear in the top beam at the centre of the span
- $M_{6,13}$ – the moment in the top beam at the centre of the span.

The results from a number of feature variations are given in Table 12.6. Note that in each case only one change from the reference model is made, as recommended in Section 2.4.4. The '% diff' columns in Table 12.6 give the percent difference from the reference model values ('refval'), i.e. % diff = (value – refvalue)/refvalue × 100. The models listed in Table 12.6 are as follows.

- *Model 2: no columns.* The columns were removed and simple supports were imposed at the lower level of the frame. This removes both rotational and horizontal restraint to the bottom beam members. Only small changes result.

Table 12.6 Sensitivity analysis models for vierendeel frame analysis

Model	Δ_{13}: mm	% diff	s_6: kN	% diff	$M_{6,13}$: kNm	% diff
1. Reference	3.539		10.658		49.02	
2. No columns	3.585	1.30	10.04	−5.80	49.21	0.39
3. Shear deformation of beams	3.815	7.80	10.37	−2.70	48.67	−0.71
4. Fixed column bases	3.493	−1.30	10.74	0.77	48.93	−0.18
5. Stiff posts	2.042	−42.30	10.17	−4.58	32.36	−33.99

- *Model 3: shear deformation of the beams (only) is included.* The deformation increases by 8%, but the effect on the internal forces is much smaller. This is the normal effect of shear deformation in a frame. It tends to affect the deflection but tends to make little difference to the internal force actions. The result of this comparison in relation to validation is that it would be best to include shear deformation (although if the software did not allow shear deformation its exclusion could be accepted).
- *Model 4: fixed column bases.* The pin supports to the columns were changed to the fully fixed condition. This makes very little difference to the chosen indicative parameters but is likely to make a significant reduction to the moments in the columns and to the behaviour under lateral load.
- *Model 5: stiff posts.* The same section as for the beams (universal beam (UB) $610 \times 305 \times 149$) was used for the posts. This makes a significant difference to the deflection and to the beam moment.

Choice of parameter variation in relation to member sizing

Model 5 of the feature variation study indicates that the span deflection is sensitive to the stiffness of the post. The question arises: 'would it be better to increase the beam stiffness or the post stiffness to stiffen the frame?'.

The deflection of the system is approximated by the equivalent beam relationship of equation (5.21) and using Table A4, and is

$$\Delta_{\text{frame}} = 5WL^3/(384E_cI_g) + WL/(8K_{\text{sv}})$$

Substituting for $I_g = A_cb^2/2$ and $K_{\text{sv}} = 24E_cI_c/(a^2(1+2\psi))$ from equations (5.16) and (5.24) and rearranging gives

$$\Delta_{\text{frame}} = \Delta_b + \Delta_{\text{sb}} + \Delta_{\text{sp}} = \frac{WL^3}{38.4EA_cb^2} + \frac{WLa^2}{192EI_c} + \frac{WLab}{96EI_p} \tag{12.1}$$

where Δ_b is the bending mode deformation due to axial deformation of the beams, Δ_{sb} is the contribution to the shear mode deformation from the beams, Δ_{sp} is the contribution to the shear mode deformation from the posts and

$W = 10.0 \times 16.0 = 160.0$ (the total load)

$\Delta_b = 160 \times 16^3/(38.4 \times (209 \times 10^6) \times 0.019 \times 3.0^2) \times 1000 = 0.478\,\text{mm}$

$\Delta_{\text{sb}} = 160 \times 16 \times 4.0^2/(192 \times (209 \times 10^6) \times 0.001259) \times 1000 = 0.811\,\text{mm}$

$\Delta_{\text{sp}} = 160 \times 16 \times 4.0 \times 3.0/(96 \times (209 \times 10^6) \times 0.000332) \times 1000 = 4.612\,\text{mm}$

$\Delta_{\text{frame}} = 0.478 + 0.811 + 4.612 = 5.901\,\text{mm}$

Element model value $\Delta_{\text{em}} = 3.539\,\text{mm}$, % diff $= -40.0\%$. The difference between the equivalent beam and the element model results is less than for the point loadcase, presumably due to the lower proportion of the total load taken as shear in the centre panels where the points of contraflexure are not close to the centre of the panels.

Taking equation 12.1 and differentiating it successively by I_c and I_p and substituting the reference model values (but with the uniformly distributed loading) gives

$$\mathrm{d}\Delta_{\mathrm{frame}}/\mathrm{d}I_{\mathrm{c}} = -WLa^2/(192EI_{\mathrm{c}}^2)$$

$$= -160 \times 16 \times 4^2/(192 \times (209 \times 10^6) \times 0.001259^2)$$

$$= -0.644\,\mathrm{m/m}^4$$

$$\mathrm{d}\Delta_{\mathrm{frame}}/\mathrm{d}I_{\mathrm{p}} = -WLab/(96EI_{\mathrm{c}}^2)$$

$$= -160 \times 16 \times 4.0 \times 3.0/(96 \times (209 \times 10^6) \times 0.00032^2)$$

$$= -14.95\,\mathrm{m/m}^4$$

It is evident that changing the I value of the posts, rather than the I value of the beams, will be the most effective way of stiffening the frame starting from the reference configuration. One should not treat this as a general result. As the value of I_{p} is increased, the post will become effectively rigid in bending, beyond which increases in the value of I_{p} will not significantly affect the frame stiffness.

12.1.8   Overall acceptance
At this point in the process there is no evidence of inadequacy in the model, but the production results have yet to be generated and assessed.

12.1.9   Modelling review document
The information included in Section 12.1 would form a basis for a modelling review document. It is likely to be considered to have too much detail in it for a conventional design but the detail would be needed in a safety critical or innovative situation.

12.2   Case study 2 – four-storey building

12.2.1   General
This example is for a 3D model of a conventional four-storey building structure.

12.2.2   Definition of the system to be modelled – the engineering model
Portrayal
Figure 12.8 shows a plan of the building, which has a steel frame with composite concrete floor slab. The slab spans 2.5 m on to secondary beams (only shown in one floor panel in Fig. 12.8). The building is braced by two cores, one of which has a row of openings in its main wall. Two bays are braced with cross diagonal members. These might not be necessary for this situation but are included to demonstrate techniques for modelling them. Foundations are pad footings on medium to stiff clay.

Details

- *Height* – four storeys at 3.7 m = 14.8 m

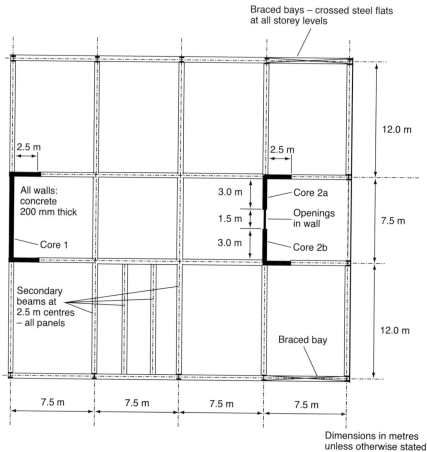

Figure 12.8 Plan of four-storey building.

- *Beams*
  - Primary beams – 7.5 m span, UB 533 × 210 × 122
  - All perimeter beams – UB 533 × 210 × 122
  - Secondary beams (apart from perimeter beams) – 12.0 m, span UB 533 × 210 × 82
- *Columns* – all columns UC (universal column) 254 × 254 × 132
- *Cores* – all walls 200 mm thick, core 2 has a row of openings with lintel beam 660 mm deep × 200 mm wide
- *Floor slabs* – 130 mm thick slabs with profiled sheeting soffits and shear connectors to beams to provide composite action
- *Diagonally braced bays* – diagonals are cross braced steel flats 150 × 20 mm

Requirements of the model
For this context it is required to estimate the displacements and internal force actions due to lateral loading.

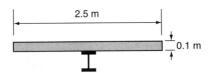

Figure 12.9 Cross section of composite beam.

12.2.3 Model development
Analysis program
The software selected for the analysis was LUSAS 13.3.[2]

Units
Units used are metres and kilonewtons.

2D or 3D model?
A 3D model may not be necessary for this analysis as simple calculations may give good estimates for most of the design variables. In this context the use of a 3D model provides a vehicle for demonstrating modelling techniques and the modelling process.

Elements and mesh
Columns
The columns are modelled as engineering beam elements with shear deformation neglected. One element is used for each storey height of a column.

Beams
The beams are modelled as engineering beam elements with shear deformation neglected. The section used is composite, as shown in Fig. 12.9.

To simplify the example, only one type of beam element section is used, based on:

- UB $533 \times 210 \times 122$ section
- concrete flange width – 2.5 m
- modular ratio – 6.0 (BS 5950: Part 3, Section 4.1) for short-term loading
- slab thickness – 100 mm (allowing for profiled shape).

The section properties for the beams quoted in Table 12.7 are for an equivalent steel section using a modular ratio of 6.0.

Floor slabs
The out-of-plane action of the floor slabs is modelled by the composite beams, which are treated as pin connected to the columns and therefore do not affect the lateral load behaviour. The in-plane action of the floor does need to be modelled. Options include:

- apply constraint equations to make the floor levels rigid in plane – this tends to be easy to implement only if there is a special facility in the program (e.g. ETABS)[3]

[2] LUSAS Finite Element Modeller, FEA Ltd.
[3] ETABS – Integrated Building Analysis and Design, Computers and Structures Inc.

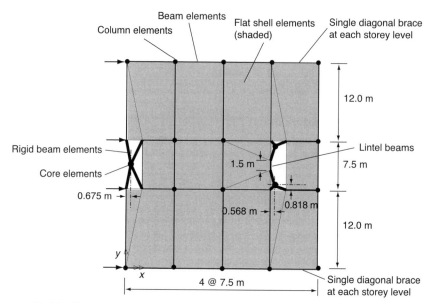

Figure 12.10 Plan of analysis model.

- use a mesh of flat shell elements
- brace the floor in-plane with axially stiff bar elements.

For visualisation of the mesh, the flat shell elements are more convenient and so are adopted for the model. The shell elements used are 130 mm thick. They are positioned at the centroidal axes of the beams and therefore the effect of their out-of-plane bending is not likely to be significant.

A single shell element was used for each floor panel, with some triangular elements to fill in at the cores (Fig. 12.10).

Cores

Options for modelling the cores include modelling them as:

- beam elements (i.e. as a column), taking account of the finite size in plan
- flat shell elements.

A main advantage of the beam element option is that the output is in the form of moments, shears and axial forces, which can be used directly for design. The output for shell elements is normally in the form of stresses. The nodal forces for a shell element mesh can be integrated to give the internal force actions, but this needs to be a feature of the software (e.g. ETABS[4]). If the core is complex, and if it is desired to model bending torsion (Section 5.4.3), then the shell element model is recommended. For this example, the beam element approach is adopted.

The core unit (or part of the core, as for core 2) is treated as a column with vertical axis at the centroid of the core part (Fig. 12.10). Rigid horizontal links (Section 5.6.2) at each storey level connect the centroids of the 'columns' to

---

[4] ETABS – Integrated Building Analysis and Design, Computers and Structures Inc.

Table 12.7 Section properties for case study 2

	Area (m²)	I_y (m⁴)	I_z (m⁴)	J (m⁴)	E (kN/m²)
All composite beams	0.0301	$1.553 \times 10^{-0.3}$	$3.388 \times 10^{-0.5}$	$1.78 \times 10^{-0.6}$	209×10^6
Columns	$1.68 \times 10^{-0.2}$	$2.25 \times 10^{-0.4}$	$7.53 \times 10^{-0.5}$	$3.19 \times 10^{-0.6}$	209×10^6
Core 1	2.5	21.1	1.56	0.033	20×10^6
Core 2 (a or b)	1.1	1.064	0.687	0.147	20×10^6
Lintel beam (core 2)	0.132	4.79×10^{-3}	0.44×10^{-3}	1.76×10^{-3}	20×10^6
Rigid beam	25	200	200	1×10^{-3}	20×10^6
Diagonal brace	0.03				209×10^6

nodes at the perimeter of the core (Section 5.6.2). In this case the rigid links are modelled using beam elements with high but finite stiffness. The main function of the rigid link is to simulate the 'plane sections remain plane' assumption for bending and therefore the bending stiffness about the axis that is in the plane of the floor slab is the important issue. The I value chosen for this model is approximately 10 times the maximum I value for core 1. The gross concrete area was used to calculate the section properties of the cores, and shear deformation is neglected.

The positions of the centroids for the cores are shown in Fig. 12.10 (these positions define the positions of the core elements in plan).

Lintel beams
It was assumed that the floor slab does not act compositely with the lintel beam and therefore the section properties used are based on the gross concrete section (660 × 200 mm). Shear deformation is neglected.

Section properties
The section properties for the model are given in Table 12.7.

Material model
Elastic behaviour is assumed.

Supports
For the columns, the BS 5950 recommendation (Section 8.2.6) of a spring that has 10% of the end rotational stiffness of the column is used, i.e.

$$K_{spring} = 0.1 \times (4EI/h)$$

where h is the column height.
The supports for the walls are assumed to be fully fixed.

Connections
The beam connections are assumed to be simply supported for bending about the horizontal axis. This is the normal assumption for design – see the validation analysis (Section 12.2.4).

Non-linear geometry

Non-linear geometry effects are, in general, neglected. One of the cross braced diagonal elements is omitted in each braced panel to account for the fact that the members are slender and unable to take compressive loading.

Loading

A lateral load of $1.2\,\text{kN/m}^2$ is applied in the global x direction.

12.2.4   Model validation

An alternative approach to validation as compared with that used in Section 12.1.5 is used here. The validation approach used is one based on risk. Risk is normally defined as the combination of the likelihood and the consequences of an event that can cause harm. Here risk is defined as the combination of the degree of uncertainty of an assumption and its importance. The levels of the uncertainty and the importance are given values (normally qualitatively assessed) in the range 1 to 5, where 5 means high uncertainty or high importance (Table 12.8). The objective is to try to ensure that no assumption falls within the shaded area of the table.

The following assumptions (A to J) were made in the analysis (needed actions are underlined).

- A: *Linear elasticity* – conventional assumption. <u>Design to codes of practice.</u>
- B: *Columns modelled as beam elements with no shear deformation* – conventional assumption. Span-to-depth ratio is 6.9:1. Shear deformation is not significant.
- C: *Beams modelled as beam elements with composite sections* – because the beams are simply supported their bending stiffnesses will not affect the behaviour under lateral load. This assumption is conservative in this context. If the beams were moment connected to the columns, accurate modelling of the behaviour would be difficult, especially in areas of hogging moment (see the sensitivity analysis, Section 12.2.6). Shear deformation is negligible because of the high span-to-depth ratio.
- D: *Cores modelled as beam elements* – this assumption would not be valid if bending torsion is important. This is unlikely to be the case in this context.

Table 12.8 Risk matrix for validation

		Importance of assumption: 5 = high				
		1	2	3	4	5
Degree of	1				B, C	
uncertainty:	2		E, F		A, D	J
5 = high	3					
	4		G, H			
	5			I		

Notes: See text for explanation of assumptions.

Shear deformation: for a wall unit, the span-to-depth ratio should be considered on a building height basis, i.e. ratios for the cores are $14.8 : 7.5 = 2.0$, $14.8 : 2.5 = 5.9$ and $14.8 : 3 = 4.9$. For the span-to-depth ratio of 2.0, shear deformation should be included. Include shear deformation for core elements.

- E: *Lintel beams as beam elements* – span-to-depth ratio is $1.5 : 0.66 = 2.3$. Include shear deformation for lintel beams.
- F: *Rigid links* – using rigid links neglects warping of the cross section, i.e. bending torsion is neglected. This is likely to be negligible in this context since the torsional stiffness of the system as a whole will be high.
- G: *Pin connections for the beams* – this will be conservative in relation to estimates of deflection under lateral load (see the sensitivity analysis, Section 12.2.6) and gives a lower bound solution for internal actions. It would not be adequate for dynamic analysis since there will be moment restraint at the beam-to-column connection.
- H: *Partial fixity for the column restraints* – conservative in this context.
- I: *Full fixity for the core restraints* – the sensitivity analysis (Section 12.2.6) shows that using an elastic spring stiffness for core 1 results in a significant redistribution of load to the bracing units. This indicates that the assumption of full fixity for the core supports may be unsatisfactory and further investigation is needed.
- J: *Non-linear geometry effect neglected* – the system is robustly braced but a check on the global critical load (Section 10.4) should be made. The appropriate compression diagonals have been removed for the loadcase shown in Fig. 12.11 (lateral load in the *x* direction). For vertical load both diagonals should be removed, although their effect on the distribution of vertical load is likely to be negligible. For lateral load in the *y* direction the compression diagonals would be different for each frame. For lateral load the choice of diagonal to remove may not cause significant differences in the results.

Outcome

Some changes are needed to the model, and the rotational fixity of the cores needs to be resolved.

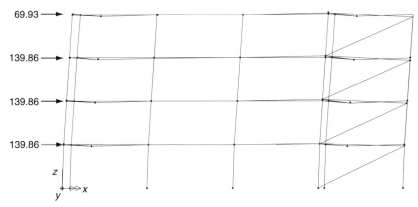

Figure 12.11 Deformed mesh for building model.

12.2.5 Results verification

The verification information in this section relates to a run with only lateral load in the x direction (Fig. 12.10).

Error warnings from software

There were no error warnings.

Data check

No errors were identified.

Sum of support reactions

- Applied load (kN): 489.510000000
- Sum of reactions: −489.5100000186

The sum of the reactions is quoted to 13 significant figures. Truncation error is low.

Symmetry check

The x direction displacements Δ_x of the structure at two symmetrical points at the top corners (nodes 400 and 458) were used for comparison. For case A in Table 12.9 the two results are only different in the last two (of 13) significant figures. This provides a check on the geometry of the structure and the loading. The displacements for case B came from an earlier run, and although the two displacements are the same for the first four significant figures they are not close enough. The fault lay in errors in defining some of the nodes. The small errors might not have been significant, but it would be unwise to continue with an analysis of this kind without providing an explanation for differences, even of this low magnitude.

Qualitative check − deformations

Figure 12.11 shows the shape of the deformed mesh for a side view looking in the y direction. Note that:

- the slope of the lateral deflection increases towards to top of the building – this is a bending mode shape consistent with the bending mode cores being the dominant source of lateral resistance
- for the diagonally braced bays, the right-hand columns are in compression (nodes deflect downwards) and the left-hand columns are in tension – consistent with the direction of loading

Table 12.9 Top deflections at symmetrical nodes

Case	Node	Δ_x
A	400	0.001284776368247
	458	0.001284776368278
B	400	0.001327660249138
	458	0.001327167061871

Table 12.10 Support reactions (base shear) in x direction

Node	F_x
Column (typical)	−0.1991
Braced frame A	−64.62
Braced frame B	−64.62
Core 1	−199.83
Core 2a	−79.12
Core 2b	−79.12

- the effect on the rotation of the rigid arms that model the rigid plane assumption for the cores is evident – the adjacent beams are displaced vertically.

There are no negative observations.

Qualitative check – force actions
When doing qualitative checks it is best to look at several features. Only two are discussed here.

Distribution of support reactions
Table 12.10 shows the main support reactions in the x direction. One would expect the following behaviour to be evident.

- Because the bending stiffness of the columns is low compared with that of the cores, the shear that they would take is likely to be low. This is confirmed by Table 12.10.
- The cores in combination are likely to take a significant proportion of the shear, with core 1 taking a bit more than cores 2a and 2b in combination because of its greater bending stiffness. This is confirmed by Table 12.10.
- The shear taken by the braced frames is likely to be less than that taken by a core, but it will be significant. This is confirmed by Table 12.10.

There are no negative observations.

Distribution of shear in the braced frames
Figure 12.12 shows the shear force in the braced frame and the applied shear on the system. The frame shear was calculated as the horizontal component of the axial

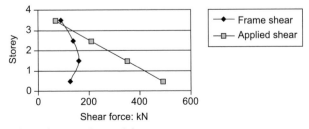

Figure 12.12 Shear force in braced frame.

force in the diagonal members. The values used in Fig. 12.12 represent the sum of the values for both braced frames.

One might predict that the distribution of shear within the height of a braced frame would have the same shape as that of the applied load. Figure 12.12 shows that such a prediction does not correspond with the results in this context. The applied shear decreases linearly with height, but the shear in the frames does not vary greatly with height. The shear in the frames at the top storey is greater than the applied shear. A prediction has been made which is not consistent with the results. Two possible reasons for this are:

- the model has errors that can be identified
- the prediction is incorrect – if this is so and if the reasons for the inconsistency can be identified then one is gaining insight into the behaviour of the system.

There is significant value in both of these situations.

In this case, it is the prediction that is incorrect. This is a good example of how examination of patterns of behaviour can promote understanding. Because the braced frames have a dominant shear mode of deformation (Section 5.10.3) they have a non-uniform interaction with the bending mode walls. The frames take proportionately more load at the top and the walls take more load at the base. The result is that often the distribution of shear on the frame is better approximated as being uniform rather than triangular. Knowing this explains the apparent anomaly in Fig. 12.12.

Checking model

Two checking models are used to estimate the top deflection and internal forces due to the x direction uniformly distributed load lateral load.

- The *rigid beam on springs* model (also known as the 'equivalent column' or 'stick' model) (MacLeod 1990) uses the top uniformly distributed (UD) load stiffnesses of the bracing units to define a top stiffness of the system. The lateral load is distributed in proportion to the top stiffness – Fig. 12.13(a).
- The *wall–frame interaction* model (MacLeod 1990) assumes that the shear mode units are represented by a spring at the top of the structure resulting

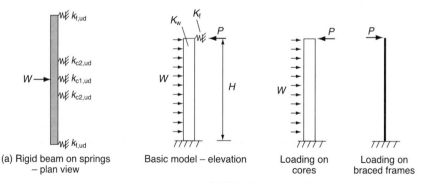

(a) Rigid beam on springs   Basic model – elevation   Loading on   Loading on
    – plan view   cores   braced frames

(b) Wall–frame interaction model

Figure 12.13 Checking models.

in the loading to them being in the form of a top point load, P, Fig. 12.13(b). Therefore, in this context it is assumed that the braced frames take only top point loading and the walls take all of the applied UD load plus a reverse point load at the top (as shown in Fig. 12.13(b)).

Total load W is calculated as

$$W = 1.2 \times 31.5 \times 14.8 = 559.4\,\text{kN}$$

Note that the total load used for the element model is 489.5 kN, being the sum of the loads shown in Fig. 12.11 (which represent the storey level loads to represent the UD load).

Note that the symbol K is used for stiffness. It generally represents a spring stiffness with units kN/m. Where a K variable has a subscript beginning with 's' it represents a shear stiffness which has units kN/radian, conforming with the definition of shear stiffness implied by equation (5.3). For springs, 'K' is used to represent a system stiffness and 'k' represents the stiffness of components.

In this section, a number of top stiffnesses are used. A top stiffness is the total lateral load to cause unit lateral displacement at the top of the unit. These are calculated using the relationships for bending or shear deflection given in Table A4. For example, from Table A4, the end deflection for a bending cantilever taking uniformly distributed load is

$$\Delta = WL^3/8EI$$

The end stiffness is therefore

$$K = W/\Delta = 8EI/L^3$$

Rigid beam on springs model

Cores The cores are treated as bending mode units with top stiffness based on UD lateral load (i.e. the lateral uniformly distributed load to cause unit top lateral deflection), hence the UD top stiffness of a core is given by

$$k_{c,ud} = 8EI/H^3$$

Core 2 is treated as two parts: core 2a and core 2b – Fig. 12.8. Therefore $E = 20 \times 10^6$, from Table 12.7: $I_{core1} = 1.558$, $I_{core2} = 0.687$, $H = 14.8$. The top UD load stiffnesses for cores 1 and 2 respectively are

$$k_{c1,ud} = 8EI/H^3 = 8 \times (20 \times 10^6) \times 1.558/14.8^3 = 76896\,\text{kN/m}$$

$$k_{c2,ud} = 8EI/H^3 = 8 \times (20 \times 10^6) \times 0.687/14.8^3 = 33907\,\text{kN/m}$$

Braced frames Figure 12.14 shows a model of a braced frame. Only the tension diagonals are included because the diagonal members are unable to take compressive loading. The braced frames are treated as equivalent beams (Section 5.10). They have a bending mode deformation component due to axial deformation of the columns and a shear mode deformation component due to the axial deformation of the diagonals.

Equation (5.22) shows that the displacement due to bending deformation and that due to shear mode deformation for an equivalent beam are added directly.

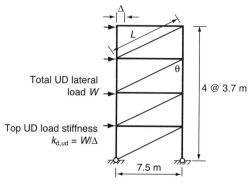

Figure 12.14 Braced frame model.

Therefore, the corresponding stiffnesses, which are reciprocals of displacement, are added reciprocally to give the total lateral stiffness; i.e. the top (UD load) stiffness of a braced frame is:

$$k_{f,ud} = 1/(1/k_{g,ud} + 1/k_{d,ud})$$

where $k_{g,ud}$ is the top (UD load) stiffness due to bending mode deformation and $k_{d,ud}$ is the corresponding stiffness due to shear mode deformation.

$$k_{g,ud} = 8EI_g/H^3 = 8 \times (209 \times 10^6) \times 0.4725/14.8^3 = 243699\,\mathrm{kN/m}$$

where

$$I_g\ (\text{equation } (5.17)) = A_c b^2/2 = 0.0168 \times 7.5^2/2 = 0.4725\,\mathrm{m^4}$$

$$k_{d,ud} = 2K_{st}/H$$

where K_{st} is the shear stiffness. (Note that the 2.0 factor in this expression is due to the UD load condition – see Table A4.)

Neglecting axial deformation of the beams and using equation (5.19)

$$K_{st} = fEA_d \sin^2 \theta \cos \theta = 1.0 \times (209 \times 10^6) \times 0.03 \times 0.8968^2 \times 0.4424$$

$$= 2231 \times 10^3\,\mathrm{kN/rad}$$

where θ is the angle between a diagonal and a column, hence

$$\cos \theta = 3.7/L = 3.7/8.363 = 0.4424$$
$$\sin \theta = 7.5/L = 7.5/8.363 = 0.8968$$
$$L = \sqrt{(3.7^2 + 7.5^2)} = 8.363\,\mathrm{m}\ (\text{see Fig. } 12.14)$$

hence

$$k_{d,ud} = 2K_{st}/H = 2 \times (2231 \times 10^3)/14.8 = 301486\,\mathrm{kN/m}$$

hence

$$k_{f,ud} = 1/(1/k_{g,ud} + 1/k_{d,ud}) = 1/(1/243699 + 1/301486) = 134765\,\mathrm{kN/m}$$

System stiffness   The top stiffness of the system is

$$K_{top,ud} = k_{c1,ud} + 2k_{c2,ud} + 2k_{f,ud}$$
$$= 76896 + 2 \times 33907 + 2 \times 134765 = 414240 \, kN/m$$

Top deflection

$$\Delta_{top} = W/K_{top,ud} = 559.4/414240 = 0.00135 \, m$$

Loads to bracing units   Distribute load in proportion to the top stiffnesses

$$\text{Load to core } 1 = Wk_{c1,ud}/K_{top} = 559.4 \times 76896/414240 = 104 \, kN$$

$$\text{Load to core 2 (both parts)} = Wk_{c2,ud}/K_{top} = 559.4 \times 2 \times 33907/414240$$
$$= 92 \, kN$$

$$\text{Load to braced frames} = Wk_{f,ud}/K_{top} = 559.4 \times 2 \times 134765/414240$$
$$= 364 \, kN$$

These are the estimations of UD load to the walls. The base shears for the walls will also have these values.

Wall–frame interaction

The basic principle for this simplified model is that the bracing units are divided into two types: those that deform predominantly in a bending mode – the walls, and those that deform predominantly in a shear mode – the frames (see Section 5.10.3 for a discussion of modes of deformation). The walls are represented by a vertical cantilever and the frames are represented by a single spring that supports the cantilever at the top (Fig. 12.13(b)). With UD lateral load on the system, the load taken by the shear mode spring P is found using

$$P = 3/8W/(1 + k_w/k_f)$$

where W is the total applied lateral UD load, k_w is the top stiffness of the combined wall units based on a top point load (i.e. top point load to cause unit top deflection) $- k_w = 3EI/H^3$ (Table A4), and k_f is the top stiffness of the shear mode units based on a top point load $- k_f = 1.0K_{st}/H$ (Table A4).

Cores   The top point load stiffness for the core units are

$$\text{Core 1 } k_{c1,p} = 3EI_{core1}/H^3 = 3 \times (20 \times 10^6) \times 1.558/14.8^3 = 28836 \, kN/m$$

$$\text{Core 2 } k_{c2,p} = 3EI_{core2}/H^3 = 3 \times (20 \times 10^6) \times 0.687/14.8^3 = 12715 \, kN/m$$

Total core top stiffness is

$$k_w = k_{c1,p} + 2k_{c2,p} = 54266 \, kN/m$$

Braced frames   The top point load stiffness of a braced frame is

$$k_{f,p} = 1/(1/k_{g,p} + 1/k_{d,p})$$

where

$$k_{g,p} = 3EI_g/H^3 = 3 \times (209 \times 10^6) \times 0.4725/14.8^3 = 91387 \, kN/m$$

Using the same value for shear stiffness as for the beam on spring model (it is unaffected by the distribution of applied load)

$$k_{d,p} = 1.0 K_{st}/H = 2231 \times 10^3/14.8 = 150743 \, \text{kN/m}$$

hence

$$k_{f,p} = 1/(1/k_{g,p} + 1/k_{d,p}) = 1/(1/91387 + 1/(150743)) = 56895$$

hence

$$k_f = 2k_{f,p} = 2 \times 56895 = 113790 \, \text{kN/m}$$

System

$$K_w/K_f = 54266/113790 = 0.477$$

$$P = 3/8 W/(1 + K_w/K_f) = 3/8 \times 559.4/(1 + 0.477) = 142 \, \text{kN}$$

Top deflection Δ_{top} is

$$\Delta_{top} = P/K_f = 142/113790 = 0.00125 \, \text{m}$$

The base shear in the walls $(W - P)$ is distributed in proportion to their bending stiffnesses

Base shear for core $1 = (W - P) \times k_{c1,p}/(k_{c1,p} + 2k_{c2,p})$

$$k_{c1,p} + 2k_{c2,p} = 28836 + 2 \times 12715 = 54266$$

Base shear for core $1 = (559.4 - 142) \times 28836/54266 = 222 \, \text{kN}$

Base shear for core 2 (both parts) $= (W - P) - 222 = (559.4 - 142) - 222$

$$= 195 \, \text{kN}$$

Summary

Table 12.11 summarises the results from the two models in comparison with the element model results.

Note that:

- both models give good correlation for top deflection
- the beam on springs model gives poor results for prediction of base shear
- the wall–frame interaction model gives satisfactory results for base shear.

Table 12.11 Summary of results from the checking models

	Component	Element model	Beam on springs model	% diff	Wall–frame interaction model	% diff
Δ_{top}: m	System	0.001288	0.001350	4.8	0.00125	-2.9
Base shear: kN	Core 1	199	104	-91.3	222	11.6
	Core 2 (×2)	178	92	-93.5	195	9.6
	Frames (×2)	129	364	64.6	142	10.1
	Total	506	560		559	

Notes: The % diff column is calculated as for Table 12.6.

One of the reasons why predictions of displacement can be more accurate than internal force actions is that the force actions are derivatives of the displacements (e.g. $M = EI\mathrm{d}^2v/\mathrm{d}x^2$) and the numerical process of calculating derivatives can cause an increase in error (whereas integration can have the reverse effect). However, the good correlation for displacement for the rigid beam model here may be mainly due to compensating assumptions. The need to look further than a single comparison is again apparent. If one only considered the deflection then one would think that the rigid beam model is good, whereas it gives very inaccurate predictions of base shears.

The wall–frame interaction model is clearly better in this context because the braced frames provide a significant degree of shear mode resistance to the lateral load.

These outcomes do not indicate any errors in the results. A similar check for loading in the y direction would be worthwhile.

12.2.6 Sensitivity analysis
Two parameters are varied in this study.

- The reference model (as used in Section 12.2.5) is changed such that the connections for the beams to the columns are moment connections (rather than pins). This results in a stiffer structure due to vierendeel action between the columns and the beams.
- The reference model is changed to provide a rotational spring support for core 1 (only) based on a medium-stiff soil. The value of the spring stiffness K_θ about the global y axis is calculated as:
  - o plan area of concrete base of wall is $9.5 \times 4.5\,\mathrm{m}$ (1.0 m wider on each side than the core itself)
  - o value of Winkler stiffness

    $k_{wk} = 50000\,\mathrm{kN/m^3}$ (Table A7)

  - o I value for foundation

    $$I_f = BD^3/12 = 9.5 \times 4.5^3/12 = 72.14\,\mathrm{m}^4$$
    $$K_\theta = k_{wk}I_f = 50000 \times 72.14 = 3.61 \times 10^6\,\mathrm{kN/m^3} \text{ (equation (8.6))}$$

- The same indicative parameters as used for the checking model (Table 12.11) are used in the results given in Table 12.12.

The following observation can be drawn from Table 12.12.

Table 12.12 Summary of sensitivity analysis results

	Component	Reference model	With beam end fixity	% diff	With support rotation	% diff
Δ_{top}: m	System	0.001286	0.001075	−16.4	0.001483	15.3
Base shear: kN	Core 1	199	197	−1.0	97	−51.3
	Core 2 (×2)	178	168	−5.6	290	62.9
	Frames (×2)	129	120	−7.0	193	49.6

- The addition of beam moment connections decreases the deflection by 16%. The base shears are not changed significantly.
- The rotational support for the core increases the deflection by 15%. The base shears are significantly different. Full fixity at the base of a core is a normal assumption, but the effect of foundation rotation on bending mode bracing units can be significant because of the magnifying effect on deflection (Section 8.2.5). Shear mode bracing units, such as the diagonally braced frames, may be less sensitive to foundation movements.

12.2.7 Model review

The validation analysis has identified the need for some minor changes and the sensitivity analysis has shown that neglecting the rotational stiffness of the supports for the shear walls can be significant. Further investigation of the latter issue needs to be carried out. An investigation of the behaviour of loading in the y direction would also be useful.

Appendix

Tables of material and geometric properties

Table A1 Areas and second moments of area of standard shapes

		Area	Second moment of area
		BD	$\frac{1}{12}BD^3$
		$BD - bd$	$\frac{1}{12}(BD^3 - bd^3)$
		$\frac{1}{2}bd$	$\frac{1}{36}bd^3$
		πR^2	$\frac{\pi}{4}R^4$
		$\pi(R^2 - r^3)$	$\frac{\pi}{4}(R^4 - r^4)$
		$2\pi Rt$	$\pi R^3 t$

Table A2 Shear areas for beams

Section shape	A_s
Rectangle	$5/6A$
Solid circle	$1.1A$
Thin-walled tube	$0.5A$
I section	$1.0A_{web}$
T section	$0.85A_{web}$

Notes: A – area of section; A_s – shear area; A_{web} – area of web.

Table A3 J values and shear stress τ for shear torsion of different cross sections

Shape	J value	Shear stress τ
Circular solid, radius R	$J =$ second polar moment of area $= \pi R^4/2$	$\tau = Tr/J$ r is radial distance
Closed thin-wall section	$J = 4A_0^2/\int \mathrm{d}s/t$	$\tau = T/(2A_0 t)$
Narrow rectangle, $b/t > 5$	$J = bt^3/3$	$\tau_{max} = 3T/bt^2$
Open section composed of narrow rectangles	$J = \sum bt^3/3$	
Solid rectangle	$J = bt^3 \left[\dfrac{1}{3} - 0.21\dfrac{t}{b}\left(1 - \dfrac{t^4}{12b^4}\right)\right]$	
Solid square	$J = 0.141b^3$	

Notes: T – applied torque; t – thickness; $\mathrm{d}s$ – differential element of length on the perimeter; A_0 – area enclosed by centreline of the wall of the section.

Table A4 Deflection formulae for beams

System (W – Total load)	Bending moment	Maximum bending deflection	Maximum shear deflection
	WL	$\dfrac{WL^3}{3EI}$	$\dfrac{WL}{A_sG}$
	$\dfrac{WL}{2}$	$\dfrac{WL^3}{8EI}$	$\dfrac{WL}{2A_sG}$
	$\dfrac{WL}{4}$	$\dfrac{WL^3}{48EI}$	$\dfrac{WL}{4A_sG}$
	$\dfrac{WL}{8}$	$\dfrac{5}{384}\dfrac{WL^3}{EI}$	$\dfrac{WL}{8A_sG}$
	$\dfrac{WL}{8}$, $\dfrac{WL}{8}$, $\dfrac{WL}{8}$	$\dfrac{WL^3}{192EI}$	$\dfrac{WL}{4A_sG}$
	$\dfrac{WL}{12}$, $\dfrac{WL}{12}$, $\dfrac{WL}{24}$	$\dfrac{WL^3}{384EI}$	$\dfrac{WL}{8A_sG}$
	$\dfrac{3}{16}WL$, $\dfrac{5}{32}WL$	$\dfrac{WL^3}{107.3EI}$	$\dfrac{WL}{4A_sG}$
	$\dfrac{WL}{8}$, $\dfrac{9}{128}WL$, $\dfrac{WL}{8}$	$\dfrac{WL^3}{185EI}$	$\dfrac{WL}{8A_sG}$
M_A ... M_B	M_A , $\dfrac{WL}{8}$, M_B	$\dfrac{5}{384}\dfrac{WL^3}{EI}-\dfrac{(M_A+M_B)L^2}{16EI}$ (at centre)	$\dfrac{WL}{8A_sG}$
	$\dfrac{WL}{6}$	$\dfrac{WL^3}{15EI}$	$\dfrac{WL}{3A_sG}$
	$\dfrac{WL}{3}$	$\dfrac{11}{60}\dfrac{WL^3}{EI}$	$\dfrac{2}{3}\dfrac{WL}{A_sG}$

Notes: for equivalent beam calculations use K_{st} (Section 5.10.4) or K_{sv} (Section 5.11.3) instead of A_sG. A_s – shear area (Table A2); G – shear modulus equation (7.6); EI – bending stiffness parameter.

Table A5 End displacement of axially loaded members

System	End displacement
	$$\Delta = \frac{NL}{EA}$$
	$$\Delta = \frac{NL}{EA\cos^2\theta}$$

Note: EA – axial stiffness parameter.

Table A6 Typical material properties for structural materials

Material	Density: kg/m^3	Young's modulus E: kN/mm^2	Poisson's ratio ν	Coefficient of linear expansion/ °C: $\times10^{-6}$
Steel	7850	209	0.3	11.5
Cast iron	7100–7500	76–145	0.3	11.0–14.0
Wrought iron	7200	169–176	0.3	10.0–12.5
Aluminium	2700	68–72	0.35	24.0
Concrete				
Normal weight	2360	20–40	0.2	10.0–14.0
Light weight	320–1920	7–19	0.2	3.6–8.0
Timber				
Softwood	500–700	5–9	0.2	4.5
Hardwood	600–1100	8–18		4.5
Brick masonry	2220	6–18		$\left\{\begin{array}{l}5.6\ (\text{hori.})\\8.0\ (\text{vert.})\end{array}\right.$
Stone masonry	2100–3000			3.0–12.0

Table A7 Typical values of Winkler stiffness for soils

Soil	k_{wk}: kN/m^3
Loose sand	4800–16000
Medium dense sand	9600–80000
Dense sand	64000–128000
Clayey medium dense sand	32000–80000
Silty medium dense sand	24000–48000
Clayey soil:	
$q_{\text{tt}} \leq 200\,\text{N/mm}^2$	12000–24000
$200 < q_{\text{tt}} \leq 400\,\text{N/mm}^2$	24000–48000
$q_{\text{tt}} > 800\,\text{N/mm}^2$	>48000

Notes: q_{tt} – bearing capacity. From Bowles (2001).

Table A8 Typical values for modulus of elasticity for soils, E_s

Soil	E_s: N/mm^2
Clay	
Very soft	2–15
Soft	5–25
Medium	15–50
Hard	50–100
Sandy	25–250
Glacial till	
Loose	10–153
Dense	144–720
Very dense	478–1440
Loess	14–57
Sand	
Silty	7–21
Loose	10–24
Dense	48–81
Sand and gravel	
Loose	48–148
Dense	96–192
Shale	144–14400
Silt	2–20

From Bowles (2001).

Table A9 Typical Poisson's ratios for soils

Type of soil	ν
Clay, saturated	0.4–0.5
Clay, unsaturated	0.1–0.3
Sandy clay	0.2–0.3
Silt	0.3–0.35
Sand (dense)	0.2–0.4
Coarse (void ratio = 0.4–0.7)	0.15
Fine grained (void ratio = 0.4–0.7)	0.25
Rock	0.1–0.4 (depends on type of rock)
Loess	0.1–0.3
Ice	0.36
Concrete	0.15

From Bowles (2001).

Bibliography

Chapter 4

The publications of NAFEMS Ltd give possibly the most comprehensive guidance available on modelling with finite elements (as distinct from the wide range of textbooks which give mainly the theory of finite elements).

Website: www.nafems.org/index.html
Mailing address: NAFEMS Ltd, Whitworth Building, Scottish Enterprise, Glasgow G75 0QD, UK, Tel: +44 13 55 22 56 88, Fax: +44 13 55 24 91 42.

For example their *Finite element primer* (1992) gives very good information about modelling with finite elements.

One of the very few textbooks on finite element analysis which focuses on modelling issues is the following.

Cook R D (1995) *Finite element modeling for stress analysis.* John Wiley.

Introductory texts include:

Chandrupatia B E (2001) *Introduction to finite elements for engineers,* international edition. Prentice Hall.
Henwood D J and Bonet J (2003) *Finite elements, a gentle introduction.* Palgrave Macmillan.

Comprehensive texts for finite element analysis include:

Cook R D, Malkus D S and Plesha M E (2001) *Concepts and applications of finite element analysis,* 4th edition. John Wiley and Sons.
Zienciewicz O C, Taylor R L and Zhu J Z (2005) *Finite element method: Its basis and fundamentals.* Butterworth-Heinemann.

Chapter 5

Structural mechanics

This is about basic theory in relation to equilibrium, stresses deformations, force–deformation relationships, etc. There are a large number of texts on this subject in print. Here is a short selection.

Bhatt P (1999) *Structures.* Prentice Hall (includes structural mechanics).
Brohn D M (2005) *Understanding structural analysis.* Stroud.
Case J, Chilver H and Ross C T F (1999) *Strength of materials and structures,* 4th edition. Butterworth-Heinemann.

Hibbeler R C (2003) *Mechanics of materials*. Pearson Prentice-Hall.
Jenkins W M (1990) *Matrix and digital computer methods in structural analysis*. McGraw-Hill Education.
Johnson D (2000) *Advanced structural mechanics*, 2nd edition, Thomas Telford.
Morgan W, Williams D, Durka F and Al Naqueim H (2002) *Structural mechanics*, in Rees D W A (ed.) (2000) *The mechanics of solids and structures*. Imperial College Press.
Smith P (2001) *An introduction to structural mechanics*. Palgrave MacMillan.
Timoshenko S P and Gere J M (2002) *Mechanics of materials*, 5th edition. Nelson Thornes.

Structural analysis and matrix structural analysis

This is about how solutions are achieved in structural analysis. A large number of texts on this subject are in print. Here is a short selection.

Bedenik B and Besant C (1999) *Analysis of engineering structures*. Horwood Publishing Ltd.
Brown T, Ghali A and Neville A (2003) *Structural analysis*, 5th edition. Spon.
Hibbeler R C (2005) *Structural analysis*. Pearson Prentice-Hall.

Modelling of connections

SCI (1995) *Modelling of steel structures for computer analysis*. Publication P148, Steel Construction Institute, Ascot.

Torsion
Shear torsion

Most texts on structural mechanics give good coverage of shear torsion theory but good information about bending torsion in English is scarce.

Bending torsion

The original development of the theory of bending torsion by Vlazov was translated into English and the following publication is still often quoted:

Vlasov V Z (1961) *Thin walled elastic beams*. Israeli Program for Scientific Translations.

Bending torsion theory is given quite good coverage in the following text.

Taranath B S (1998) *Steel, concrete and composite design of tall buildings*, 2nd edition. McGraw-Hill Education.

The following texts cover bending torsion but are now out of print:

Murray N W (1986) *Introduction to the theory of thin walled structures*. Clarendon Press.
Zbhirohowski-Koscia K (1967) *Thin walled beams*. Crosby Lockwood.

Non-linear modelling

Bazant Z and Jirásek M (2001) *Inelastic analysis of structures*. Wiley.
Maekawa K, Okamura H and Pimanmas A (2003) *Non-linear mechanics of reinforced concrete*. Spon Press.

Chapter 8
Finite element modelling in geotechnical engineering

Muir Wood D (2004) *Geotechnical modelling*. Spon Press.

Potts D and Zdravkovic L (1999) *Finite element analysis in geotechnical engineering: Volume I – Theory*. Thomas Telford.

Potts D and Zdravkovic L (2001) *Finite element analysis in geotechnical engineering: Volume II – Applications*. Thomas Telford.

Boundary elements

Lebeda A, Hall W S and Oliveto G (2003) *Boundary element methods for soil–structure interaction*. Kluwer Academic Publishers.

Foundations

Bowles J E (2001) *Foundation analysis and design*, 5th edition. McGraw-Hill Education.

Rafts

Hemsley J A (1998) *Elastic analysis of raft foundation*. Thomas Telford.

Hemsley J A (ed.) (2000) *Design applications of raft foundations*. Thomas Telford.

Poulos H G and Davis E H (1980) *Pile foundation analysis and design*. John Wiley & Sons Inc.

Wood L A (1978) RAFTS: A program for the analysis of soil structure interaction, *Adv. Eng. Software* **1**, No 1, 11.

Piling

Fleming W G K, Weltman A J and Randolph M F (1994) *Piling engineering*. Taylor and Francis.

Chapter 9

Wind loading

Holms J D (2001) *Wind loading of structures*. Spon Press.

Smith P D and Hetherington J G (1994) *Blast and ballistic loading of structures*. Butterworth-Heinemann.

Taranath B (2004) *Wind and earthquake resistant buildings: structural analysis and design*. CRC Press.

Earthquake loading

Dowrick D J (2003) *Earthquake risk reduction*. John Wiley and Sons.

Gupta A K (1990) *Response spectrum method in seismic analysis and design of structure*. CRC Press.

Paulay T and Priestly M J N (1992) *Seismic design of reinforced concrete and masonry buildings*. Wiley.

Scarlat A S (1996) *Approximate methods in structural seismic design*. Spon.

Williams A (2001) *Seismic design of buildings and bridges*. Oxford University Press.

Chapter 10

Cable systems

Broughton P and Ndumbaro P (1994) *Analysis of cable and catenary structures*. Thomas Telford.

Buchholdt H A (1998) *An introduction to cable roof systems*. Thomas Telford.

BIBLIOGRAPHY   179
Elastic stability

Kollár L (1999) *Structural stability in engineering practice*. E & FN Spon.

Wang C M, Wang C Y and Reddy J N (2004) *Exact solutions for buckling of structural members*. CRC Press.

Zalka K (2000) *Global structural analysis of buildings*. E & FN Spon.

Chapter 11

Bhatt P (2002) *Programming the dynamic analysis of structures*. Spon Press.

Buchholdt H A (1997) *Structural dynamics for engineers*. Thomas Telford.

Chopra A K (2000) *Dynamics of structures – theory and application to earthquake engineering*. Prentice Hall.

Clough R W and Penzien J (1993) *Dynamics of structures*. McGraw-Hill Education.

Humar J (2002) *Dynamics of structures*, 2nd edition. Balkema.

Kappos A J (2002) *Dynamic loading and design of structures*. Spon Press.

Maguire J R and Wyatt T A (1999) *Dynamics – An introduction for civil and structural engineers*. Thomas Telford.

References

ACI (2002) *Suggested analysis and design procedures for combined footings and mats.* American Concrete Institute 336.2R-88 (reapproved 2002).

Ades N and MacLeod I A (1992) Computer aided design of reinforcement for concrete slabs under biaxial bending. *Structural Engineering*, **4**, No 1, 73–79.

AGS (1994) *Guide to the validation and use of geotechnical software.* Association of Geotechnical Specialists, Camberley, Surrey.

Argyris J H (1954) Energy theorems and structural analysis. *Aircraft Engineering*, **26**, 137–383.

Baker J F, Horne M R and Heyman J (1951) *The Steel Skeleton.* Cambridge University Press.

Bowles J E (2001) *Foundation analysis and design*, 5th edition. McGraw-Hill Education.

BSI (1997) *BS 6399: Loading for buildings: Part 2, Wind Loading.* British Standards Institution.

BSI (2000) *BS 5950-1: Structural use of steelwork in building. Code of practice for design – rolled and welded sections.* British Standards Institution.

CEN (2004a) *BS EN 1992-1-1:2004 Eurocode 2: Design of concrete structures. General rules and rules for buildings.* British Standards Institution.

CEN (2004b) *BS EN 1993-1-1:2004 Eurocode 3: Design of steel structures. General rules and rules for buildings.* British Standards Institution.

Clark L A (1976) The provision of and compression reinforcement to resist in-plane forces. *Magazine of Concrete Research*, **8**, No 94, 3–12.

Davies J M (2001) Second-order elastic–plastic analysis of plane frames, in Zingoni A (ed.) *Structural Engineering, Mechanics and Computation*, Proceedings of the International Conference on Structural Engineering Mechanics and Computation, Cape Town, 2–4 April 2001, Elsevier.

Desai C S and Abel J F (1972) *Introduction to the finite element method.* Van Nostrand Reinhold.

Foeroyvik F (1991) The Sleipner accident – tricell calculation and reinforcement error. *Finite Element News*, No 6, 27–29.

Ghali A and Favre R (2002) *Concrete structures; stresses and deformations.* Spon Press.

Gordon S R and May I M (2004) Observations on the grillage analysis of slabs. *The Structural Engineer*, **82**, No 3, 35–38.

Griffiths H, Pugsley A G and Saunders O (1968) *Report of the inquiry into the collapse of flats at Ronan Point, Canning Town.* Her Majesty's Stationery Office.

Hambly E (1991) *Bridge deck behaviour*, 2nd edition. E & F N Spon.

Hemsley J A (1998) *Elastic analysis of raft foundation.* Thomas Telford.

Irons B and Ahmad S (1980) *Techniques of finite elements.* Ellis Horwood.

ISO 9001 (2000) *Quality Systems Part 1: Specification for design/development, production installation and servicing.* International Organization for Standardization.

IStructE (1978) *Structure–soil interaction: A state of the art report.* Institution of Structural Engineers.

IStructE (1989) *Soil–structure interaction: The real behaviour of structures.* Institution of Structural Engineers.

IStructE (2002) *Guidelines for the use of computers for engineering calculations.* Institution of Structural Engineers.

Levy M and Salvadori M (1992) *Why buildings fall down.* W W Norton & Co.

MacLeod I A (1967) Lateral stiffness of shear walls with openings, in Coull A and Stafford Smith B (eds) *Tall Buildings*, Pergamon.

MacLeod I A (1987) Use of bed-joint reinforcement in brick buildings to resist settlement. *The Structural Engineer*, **65A**, No 10, 369–376.

MacLeod I A (1990) *Analytical modelling of structural systems.* Ellis Horwood.

MacLeod I A (1995) A strategy for the use of computers in structural engineering. *The Structural Engineer*, **73**, No 1, 366–370.

MacLeod I A and Zalka K A (1996) The global critical load ratio approach to stability of building structures. *The Structural Engineer*, **74**, No 24, 405–411.

Martin T and MacLeod I A (1995) The Tay Rail Bridge disaster – a reappraisal based on modern analysis methods. *Proceeding of the Institution of Civil Engineers*, **108**, No 2.

McGuire W, Gallagher R W and Ziemian R D (2000) *Matrix structural analysis.* John Wiley.

Moy S J (1996) *Plastic methods for steel and concrete structures*, 2nd edition. MacMillan.

NAFEMS (1990) *The standard NAFEMS benchmarks.* NAFEMS Ltd (see bibliography Chapter 4).

NAFEMS (1992) *A finite element primer.* NAFEMS Ltd.

NAFEMS (1995) *SAFESA management guidelines.* NAFEMS Ltd.

NAFEMS (1999) *QSS to BS EN ISO 9001 relating to engineering analysis in the design and integrity demonstration of engineered products.* NAFEMS Ltd.

O'Brien E J and Keogh D L (1999) *Bridge deck analysis.* Routledge.

Priestly M J N (2003) *Myths and fallacies in earthquake engineering, revisited.* IUSS Press.

Straub H (1964) *A history of civil engineering.* MIT Press.

Taranath B S (1998) *Steel, concrete and composite design of tall buildings*, 2nd edition. McGraw-Hill Education.

Timoshenko S P and Gere J M (1961) *Theory of elastic stability.* McGraw-Hill Education.

Timoshenko S P and Woinowsky-Krieger S (1964) *Theory of plates and shells.* McGraw-Hill.

Trahair N S (1993) *Flexural-torsional buckling of structures: new directions in civil engineering.* CRC Press.

Troitsky M S (1967) *Orthotropic bridges: theory and design.* James F Lincoln Arc Welding Foundation.

Turner M J, Clough R W, Martin H C and Topp L J (1956) Stiffness and deflection analysis of complex structures. *Journal of Aeronautical Science*, **23**, 805–823.

Vlasov V Z (1961) *Thin walled elastic beams.* Israeli Program for Scientific Translations.

Whittle R T (1985) *Design of reinforced concrete flat slabs to BS8110*, CIRIA Report 110, CIRIA.

Wood R H (1969) The reinforcement of slabs in accordance with a predetermined field of moments. *Concrete*, **2**, No 2, 69–79.

Wyatt T A (1988) *Design guide on the vibration of floors.* Steel Construction Institute.

Young W C and Budynas R G (2001) *Roark's formulas for stress and strain*, 7th edition. McGraw-Hill.

Zalka K (2002) Buckling analysis of buildings braced by frameworks, shear walls and cores. *Structural Design of Tall Buildings*, **11**, 197–219.

Zienciewicz O C and Cheung Y K (1965) Finite elements in the solution of field problems. *The Engineer*, 24 Sept, 507–510.

Index

Page numbers in *italics* refer to diagrams or illustrations